ModelMate—A Graphical User Interface for Model Analysis

By Edward R. Banta

Chapter 4 of
Section E, Model Analysis
Book 6, Modeling Techniques

Groundwater Resources Program

Techniques and Methods 6–E4

U.S. Department of the Interior
U.S. Geological Survey

U.S. Department of the Interior
KEN SALAZAR, Secretary

U.S. Geological Survey
Marcia K. McNutt, Director

U.S. Geological Survey, Reston, Virginia: 2011

For more information on the USGS—the Federal source for science about the Earth, its natural and living resources, natural hazards, and the environment, visit http://www.usgs.gov or call 1-888-ASK-USGS

For an overview of USGS information products, including maps, imagery, and publications, visit http://www.usgs.gov/pubprod

To order this and other USGS information products, visit http://store.usgs.gov

Suggested citation:
Banta, E.R., 2011, ModelMate—A graphical user interface for model analysis: U.S. Geological Survey Techniques and Methods 6–E4, 31 p.

Preface

This report documents ModelMate, a graphical user interface for model analysis. The documentation presented herein describes installation and use of the application. The performance of ModelMate has been successfully tested in a variety of modeling and model-analysis scenarios. If, however, errors are detected in this document or in the functioning of the program, users are requested to notify the U.S. Geological Survey. Updates might occasionally be made to this document and to the program. Users can download the software and check for updates on the Internet at *http://water.usgs.gov/software/ModelMate/*.

Contents

Figures

Tables

ModelMate—A Graphical User Interface for Model Analysis

By Edward R. Banta

Abstract

ModelMate is a graphical user interface designed to facilitate use of model-analysis programs with models. This initial version of ModelMate supports one model-analysis program, UCODE_2005, and one model software program, MODFLOW-2005. ModelMate can be used to prepare input files for UCODE_2005, run UCODE_2005, and display analysis results. A link to the GW_Chart graphing program facilitates visual interpretation of results. ModelMate includes capabilities for organizing directories used with the parallel-processing capabilities of UCODE_2005 and for maintaining files in those directories to be identical to a set of files in a master directory. ModelMate can be used on its own or in conjunction with ModelMuse, a graphical user interface for MODFLOW-2005 and PHAST.

Introduction

ModelMate is a graphical user interface (GUI) designed to facilitate model analysis. The GUI provides an intuitive visual layout of the various controls and capabilities commonly used by a model-analysis program. When suitably populated with information related to the model and the model-analysis program, ModelMate can be used to prepare input for the model-analysis program, run it, import results of the program, and perform related tasks. The actual analyses are performed by the model-analysis program, which is separate from the ModelMate program. UCODE can perform the following analyses: sensitivity analysis, parameter estimation, tests of model linearity, prediction sensitivity, nonlinear uncertainty, and investigation of an objective function (Poeter and others, 2005). ModelMate can be used to prepare input required by UCODE to perform these analyses, and it can then invoke UCODE. After an analysis, ModelMate can be used to invoke GW_Chart (Winston, 2000) to graphically display analysis results.

Purpose and Scope

This report documents ModelMate, a GUI designed to facilitate use of model-analysis programs with models. In this initial version of ModelMate, one model-analysis program (UCODE_2005, Poeter and others, 2005) and one model software program (MODFLOW-2005, Harbaugh, 2005) are

explicitly supported. In the rest of this report and in most places in ModelMate, UCODE_2005 is referred to as simply "UCODE." Although the initial version specifically supports only UCODE as the model-analysis program, ModelMate is designed to accommodate addition of support for other model-analysis programs. Similarly, support could be added for model software other than MODFLOW-2005.

This documentation describes the use of ModelMate to prepare input for UCODE, to run UCODE, and to perform related tasks. Although this document references analyses supported by UCODE, documentation of these analyses is beyond the scope of this report; the reader is referred to Poeter and others (2005) for complete discussion of the analyses. Model-calibration methods and related concepts are described in detail by Hill and Tiedeman (2007).

Installation and Setup

ModelMate can be obtained from the Internet address provided in the Preface. The distribution file is an executable archive file. When downloaded and executed on a computer running the Windows operating system, the distribution file creates a directory (folder) containing subdirectories and files required for running ModelMate. The subdirectories are named Bin, Completed_Example, Doc, and Example.

The Bin subdirectory contains the executable files needed by ModelMate. ModelMate can be invoked by starting the executable file ModelMate.exe. If ModelMate.exe is copied to another directory, "accjupiter.dll" also needs to be copied to the same directory. The Doc subdirectory contains Portable Document Format (PDF) files containing this report and a tutorial. The tutorial provides step-by-step instructions intended for first-time users. The Example subdirectory contains files referenced in the tutorial. The Completed_Example directory contains files as they would be on completion of the tutorial.

A UCODE executable file is distributed with ModelMate; however, when new versions of UCODE are released, an updated executable file of UCODE can be used instead. No modification of UCODE is required to work with ModelMate. As described in this report, UCODE is used to invoke MODFLOW-2005. A MODFLOW-2005 executable file also is distributed with ModelMate. When updated versions of MODFLOW-2005 are released, the new executable file can be used instead. No modification of MODFLOW-2005 is required to work with UCODE or ModelMate.

Conventions

In using the term "model," a convention is needed to distinguish between (a) the software, and (b) a combination of the software and input files developed to represent and simulate a process in either a hypothetical or a real-world system. In this report, the term *model software* is used to refer to the software, which generally means an executable file on a computer, but it also may refer to the source code used to generate the executable file. MODFLOW-2005 (Harbaugh, 2005) is an example of model software. The term *model* in this report refers to a combination of a model executable file and input files required to enable the model software to simulate a user-defined system of interest.

A number of capabilities of ModelMate are accessed by menus near the top of various windows. In this documentation, names of menus and menu items are designated by the use of a bold font and the pipe ("|") symbol. So, for example, **File|New Project** means to open the "File" menu and select menu item "New Project."

In UCODE documentation, the term *directory* is used to denote a file-system object referred to as a *folder* in the Windows operating system. These terms are used interchangeably in this report, depending on the context.

Vocabulary

This section lists a number of terms used throughout the documentation. Understanding of the italicized terms in this section is essential to successful use of ModelMate. Users who are unfamiliar with these terms are encouraged to refer to Hill and Tiedeman (2007) or Poeter and others (2005).

Process simulation models read input, perform calculations that simulate one or more processes, and write output. The processes involved may be physical, as in MODFLOW; chemical, as in PHREEQC (Parkhurst and Appelo, 1999); or any other simulated process. For example, economic models could be considered process simulation models.

Selected numeric model inputs generally are termed model *parameters*. In ModelMate, only real, continuous variables can be treated as parameters; methods for discrete variables are not included.

Numeric model outputs include *simulated values* that may be compared to measured (or "observed") values and simulated values that may be considered *predictions* of future conditions. Observed values commonly are referred to as "*observations*," and the corresponding model-simulated values are referred to as "simulated equivalents to observations." Only continuous real values can be considered by the version of ModelMate documented in this report.

Model calibration is a process whereby various aspects of process-model construction and parameter values are modified in hopes of improving how well the model represents the system of concern. Part of model calibration includes changing the values of selected parameters used to define model inputs. This part of the model calibration process is amenable to application of optimization, and is the process that is served by ModelMate.

Parameter estimation is the process of adjusting model parameters, within reasonable limits, such that simulated equivalents are as close as possible to observed values. The difference between an observed value and a simulated value is a *residual*. *Observation error* is taken into account by assigning *weights* to each observed value. *Model fit* commonly is characterized numerically by an *objective function*, which is an expression that combines contributions from all residuals and their weights. A goal of parameter estimation is to minimize the numeric value of a specified objective function.

Parameter estimation can be performed by manual trial-and-error adjustment of model parameters, by any of a number of software-encoded algorithms, or by some combination of methods. In this document parameter estimation refers to adjusting parameter values using a software algorithm.

Field information related to model parameters can be used in parameter estimation to provide a preference for a specified numeric value for a model parameter; when used in parameter estimation, this type of information is called *prior information*. As with observed values, prior information can be assigned a weight based on an analysis of errors in the prior information. The objective function may include terms involving prior-information items and their weights.

In this documentation, *sensitivity* refers to the degree to which a model-simulated value depends on a model parameter; specifically, sensitivity is the partial derivative of a model-simulated value with respect to a parameter. Sensitivity commonly is approximated by a finite-difference technique involving *perturbation* of parameters. UCODE and programs bundled with it can perform various types of *residual, uncertainty,* and *linearity analyses*. The initial version of ModelMate can invoke two types of residual analysis. To perform the uncertainty and linearity analyses, the user will need to work outside of ModelMate.

Design Overview

ModelMate is a GUI designed to facilitate use of model-analysis programs and models. One model-analysis program, UCODE (Poeter and others, 2005), is supported, and one model software program, MODFLOW-2005 (Harbaugh, 2005), is supported. UCODE supports several types of model analyses, including parameter estimation and sensitivity analysis, among other analyses. UCODE is designed to perform these analyses on a functioning process-simulation model. For any model analysis to be successful, the user needs to have a model-software executable file and appropriate model input files that enable the model software to run a simulation to completion. The model may represent a hypothetical system or a real-world system. The model can be developed by manual preparation of input files or by preprocessing software designed specifically for the model software being used. For example, the MODFLOW GUI (Winston, 2000), ModelMuse (Winston, 2009), or MFI2005 (Harbaugh, 2010) can be used to prepare input files for a model based on MODFLOW-2005. The MODFLOW input files prepared by any of these programs can then be used with UCODE and ModelMate.

ModelMate provides a visual representation of the model-analysis process. A useful way to view this process is to categorize some aspects as being controls related to the model analysis and other aspects as relating to the connection between the model-analysis application and the model. This distinction is particularly useful because the connection between UCODE and the model is partly based on template files (for generating model input files) and instruction files (for extracting simulated values from model output files) that conform to specifications of the JUPITER Application Programming Interface (API) (Banta and others, 2006). The JUPITER API is a set of Fortran-90 modules—including data, data types, subroutines, and functions—designed to facilitate construction of model-analysis programs by providing commonly needed capabilities. Any model-analysis application that uses the JUPITER API for this connection will be a potential candidate for inclusion among the model-analysis applications supported by future versions of ModelMate, with minimal need to make modifications to accommodate the application/model connection for the new application.

Other aspects of the model-analysis application/model connection include the definitions of parameters, observations, and predictions. UCODE allows the user substantial flexibility in defining these items; ModelMate allows the user to configure its interface to provide similar flexibility. UCODE uses the JUPITER API to read the data to define these items, and ModelMate prepares these data accordingly. Again, the potential exists to add support in ModelMate for a model-analysis application that uses the JUPITER API to read data related to parameters, observations, and(or) predictions.

Model analysis commonly requires numerous simulations of the model being analyzed. In many cases, simulations are independent of each other, and parallelization of the simulations, if supported by the model-analysis application, is appropriate. UCODE supports parallelization of simulations involved in calculation of parameter sensitivities by perturbation. ModelMate constructs UCODE input related to parallelization and facilitates copying of files to directories involved in parallel processing.

Once ModelMate is populated with the information needed by UCODE for a particular model analysis, it can be used to create UCODE input files, invoke UCODE, and view UCODE output. ModelMate also provides ways to invoke various postprocessors for analyzing and graphing UCODE output.

ModelMate and ModelMuse are designed to work together. Menu items in ModelMuse can be used to prepare a partly populated ModelMate project file, and updated parameter values resulting from a UCODE parameter-estimation analysis can be communicated back to ModelMuse through an updated ModelMate project file.

Projects and Project Files

ModelMate stores all information required for a model-analysis project in a ModelMate project file (or "ModelMate file"), which is a text file with the extension "mtc" in a format readable by ModelMate. To ensure that the file remains readable by ModelMate, users generally should avoid editing a ModelMate file with a text editor.

Running ModelMate

ModelMate can be invoked in a number of ways. A shortcut to the ModelMate.exe executable file can be placed on the Windows desktop; opening the shortcut will then start ModelMate. Alternatively, Windows Explorer can be used to navigate to the directory where ModelMate.exe and accjupiter.dll reside. Opening ModelMate.exe will then start ModelMate. ModelMate and ModelMuse (Winston, 2009) are designed to work together. ModelMate can be invoked from ModelMuse, as described in the section titled "Use of ModelMate with ModelMuse."

When ModelMate is started by opening a desktop icon pointing to ModelMate.exe or by opening ModelMate.exe in Windows Explorer, an initial "Select ModelMate Project" dialog window appears, asking the user to open an existing ModelMate project file or to select a directory and specify a name for a new project file. When creating a new ModelMate Project, a filename without the ".mtc" can be specified; the extension will be added automatically.

Users may find it convenient to use Windows to associate the extension "mtc" with the ModelMate.exe executable. When this association is made, ModelMate can be started by opening any file with the extension "mtc" in Windows Explorer. ModelMate will then attempt to open the selected file, and the introductory "Select ModelMate Project" dialog will be skipped.

Interface Elements

The main window of ModelMate as it appears at startup is shown in figure 1. The blue bar at the top is the title bar, and it displays the full pathname of the current ModelMate file if the title bar is wide enough to accommodate it; if the full pathname is too long, only the filename is displayed. Below the title bar is the menu bar, showing menus labeled **File**, **Project**, **UCODE**, **Model**, and so forth. Clicking on a menu will display a list of menu items and(or) submenus. The appendix provides a summary of all menu items. The tool bar, located below the menu bar, provides quick access to several of the items on the menus. When the mouse pointer is allowed to hover over a button on the tool bar, a hint will be displayed, identifying the action that will take place when the button is clicked. Below the tool bar are two panels, labeled "Model-Analysis Application" and "Application/Model Connection." At the bottom of the main ModelMate window is a status bar with two sides. The left side displays the user-assigned project name, and the right side is used to display hints. Clicking on the status bar displays the hint in a dialog box.

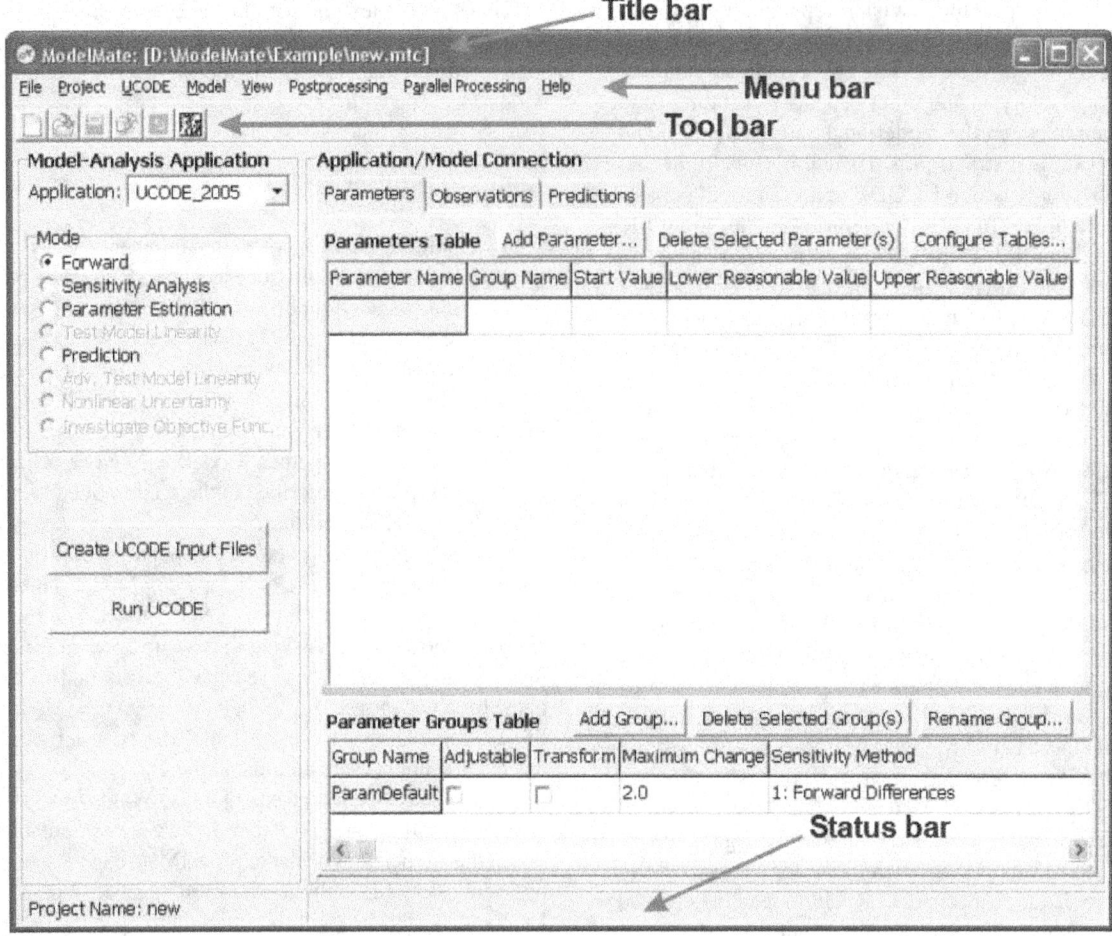

Figure 1.   Main ModelMate window at startup.

Model-Analysis Applications

ModelMate supports the model-analysis application UCODE (Poeter and others, 2005). Addition of support for other model-analysis applications is anticipated as warranted by demand and as time and resources allow. Users can check the URL in the Preface for updates, which may add support for other model-analysis applications.

At the top of the "Model-Analysis Application" panel is a drop-down list labeled "Application." This control identifies the active application. In the initial version of ModelMate, only UCODE is supported, and the drop-down list contains only "UCODE_2005." Support for additional applications may be added in future versions of ModelMate, however, and this drop-down list will enable the user to select the active application. If multiple applications are supported, this selection will affect the controls visible in the "Model-Analysis Application" panel and the list of menus in the menu bar.

Models

ModelMate explicitly supports the model software MODFLOW-2005 (Harbaugh, 2005). Addition of support for other model software is anticipated as warranted by demand and as time and resources allow. Users can check the URL in the Preface for updates, which may include support for other model software.

Users may find ModelMate to be useful in using UCODE to analyze models based on model software that is not explicitly supported. In this situation, users would need to prepare template and extraction-instruction files (see the section titled "Model Input and Output Files" in this report and Poeter and others [2005]) appropriate for the model software. When those files have been prepared, ModelMate can be used much as described in this report to prepare input files for UCODE, to invoke UCODE, and to invoke postprocessors to analyze the results.

Each ModelMate project can store information about one or two models. Typically, one model can be considered a primary model; when parameter estimation is used, parameters of the primary model are estimated. In modeling practice, it is common to have one model that simulates a calibration period and a second model that simulates a predictive period. In this situation, the calibration-period model would be the primary model. In other cases, a single model may fulfill the needs of the modeler, and there would be no predictive model. In the ModelMate interface, data and settings associated with the predictive model are clearly labeled as such. Where labels do not indicate an association with the predictive model, data and settings apply to the primary model or to the ModelMate project as a whole.

Program Locations

ModelMate allows the user to directly invoke a number of external programs; in addition, it uses the location of the MODFLOW-2005 executable file in generating model commands to be invoked by UCODE. Locations of program executable files, for example ucode_2005.exe (UCODE_2005) and mf2005.exe (MODFLOW-2005), are stored by ModelMate under the user's "Documents and Settings" folder in a file named ModelMate.ini under Application Data\WRDAPP\ModelMate.

To set or change pathnames to program executable files, use menu item **Project|Program Locations** (appendix) to open the "Program Locations" window. Ensure that paths to the UCODE and MODFLOW-2005 executable files are valid. If any path listed on the "Program Locations" window is invalid, its edit box will be red. For convenience, executable files for MODFLOW-2005 (mf2005.exe) and UCODE (ucode_2005.exe) are provided in the bin directory of the ModelMate distribution. When newer versions of these programs are released, the "Program Locations" window can be used to update the ModelMate project to use the new locations. The "Program Locations" window also allows the user to store pathnames for the postprocessor graphing program GW_Chart (Winston, 2000) and two postprocessors, Residual_analysis and Residual_analysis_adv, which are bundled with UCODE.

File Names

All file names—for example, input and output files for the model and the model-analysis application—are stored as relative pathnames in a ModelMate project file. The file names are displayed as absolute pathnames, but they are stored internally as relative pathnames. This design allows a directory containing a ModelMate project, possibly containing subdirectories with related files, to be moved as a unit to another directory or another computer. When this is done, all files should remain accessible to ModelMate, UCODE, and MODFLOW-2005.

Model command lines are stored exactly as displayed in the "Model Commands" window (menu item **Model|Commands to Invoke Model**). The command lines can be edited or recreated from the relative pathnames as needed if directories are moved or reorganized.

Variable Naming Convention

User-defined names for parameters, observations, predictions, prior-information items, and all groups are required to conform with the JUPITER API naming convention (Banta and others, 2006). The following two rules constitute the naming convention:

Rule 1. The first character needs to be a letter of the English alphabet.

Rule 2. All characters after the first letter need to be a letter, digit, or member of the set: "_", ".", ":", "&", "#", "@" (underscore, dot, colon, ampersand, number sign, at symbol).

Names of parameters and all groups are limited to 12 characters. Names of observations, predictions, and prior-information items are limited to 20 characters.

File Viewer

ModelMate includes a file viewer, which is used to view UCODE input and output files. The file viewer provides access to two windows—one for viewing the main UCODE input file and the other for viewing UCODE or postprocessor output files. The file viewer is activated from the **View** menu by selecting either **View|UCODE Main Input File** or **View|UCODE Main Output File**. The file viewer also is activated when either **Postprocessing|Residual_analysis** or **Postprocessing|Residual_analysis_adv** is invoked; in these cases, the file viewer displays the main output file generated by the selected postprocessing application.

The file viewer is illustrated in figure 2. The title bar displays the name of the file. The menu bar has three menus, **File**, **Navigation**, and **View**. Each menu option also may be accessed by a button on the tool bar with the associated icon or by a keyboard shortcut. Table 1 describes these options.

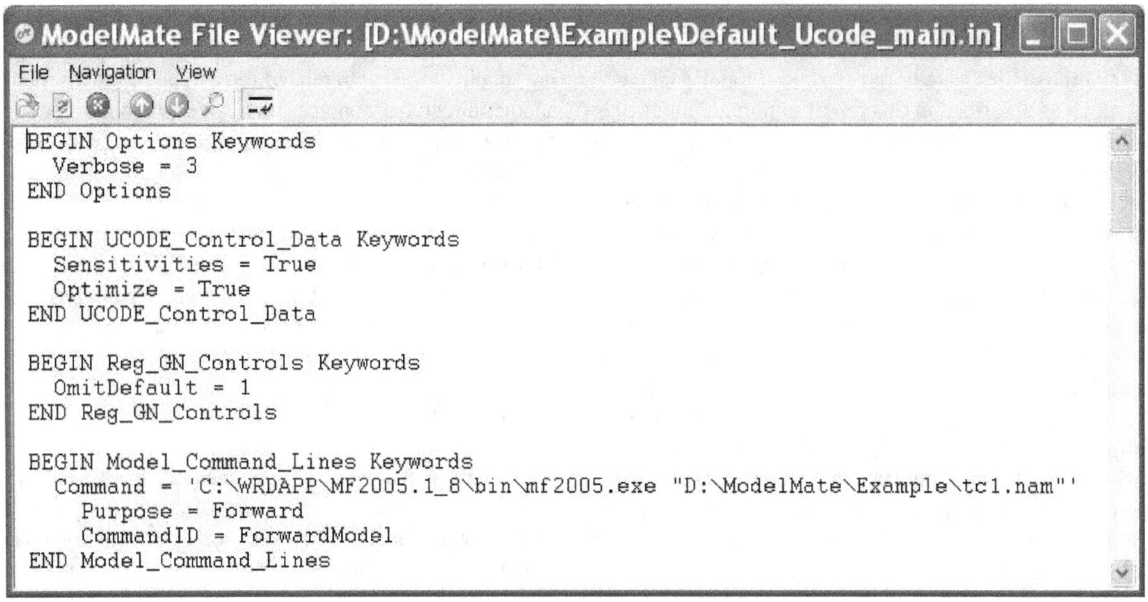

```
⊘ ModelMate File Viewer: [D:\ModelMate\Example\Default_Ucode_main.in]  _ □ X
File  Navigation  View
⟳ ▣ ⊗ ⊙ ⊙ ⌕ ⇥

BEGIN Options Keywords
   Verbose = 3
END Options

BEGIN UCODE_Control_Data Keywords
   Sensitivities = True
   Optimize = True
END UCODE_Control_Data

BEGIN Reg_GN_Controls Keywords
   OmitDefault = 1
END Reg_GN_Controls

BEGIN Model_Command_Lines Keywords
   Command = 'C:\WRDAPP\MF2005.1_8\bin\mf2005.exe "D:\ModelMate\Example\tc1.nam"'
      Purpose = Forward
      CommandID = ForwardModel
END Model_Command_Lines
```

Figure 2. File viewer.

Table 1. File viewer menu items.

Menu	Menu item	Shortcut	Action
File	⟳ Open	\<Ctrl\>-O	Browse to open any text file
	▣ Reload current file	\<Ctrl\>-R	Reload the file named in the title bar
	⊗ Exit	\<Ctrl\>-E	Close the file viewer window
Navigation	⬆ Top	\<Ctrl\>-T	Go to top of file
	⬇ Bottom	\<Ctrl\>-B	Go to bottom of file
	⌕ Find	\<Ctrl\>-F	Find text in file
View	⇥ Word wrap	\<Ctrl\>-W	Toggle word wrapping

Basic Tasks

Table Configuration for Members and Groups

Substantial flexibility is supported by UCODE in defining parameters, observations, predictions, and prior information. In each case, users can define groups for convenience of input and for analysis by groups. ModelMate supports this flexibility by allowing the user to configure tables related to parameters, observations, predictions, and prior information as desired, so that attributes can be assigned by individual item, by group, or by default, within the limitations of UCODE input requirements. Each window containing tables that can be configured in this way has a button labeled "Configure Tables" (fig. 1). The "Configure Tables" button allows access to a dialog window that can be used to configure the tables in that window. For the Observations and Observation Groups tables, the "Configure Tables" button provides access to the "Configure Observation Tables" window (fig. 3). For each attribute, a choice of at least two of the options "Observations Table," "Observation Groups Table," and "No Table (use default)" is provided. Clicking in the cell in the "Select Table or Default" column for the attribute to be changed brings up a drop-down list of valid options for that attribute.

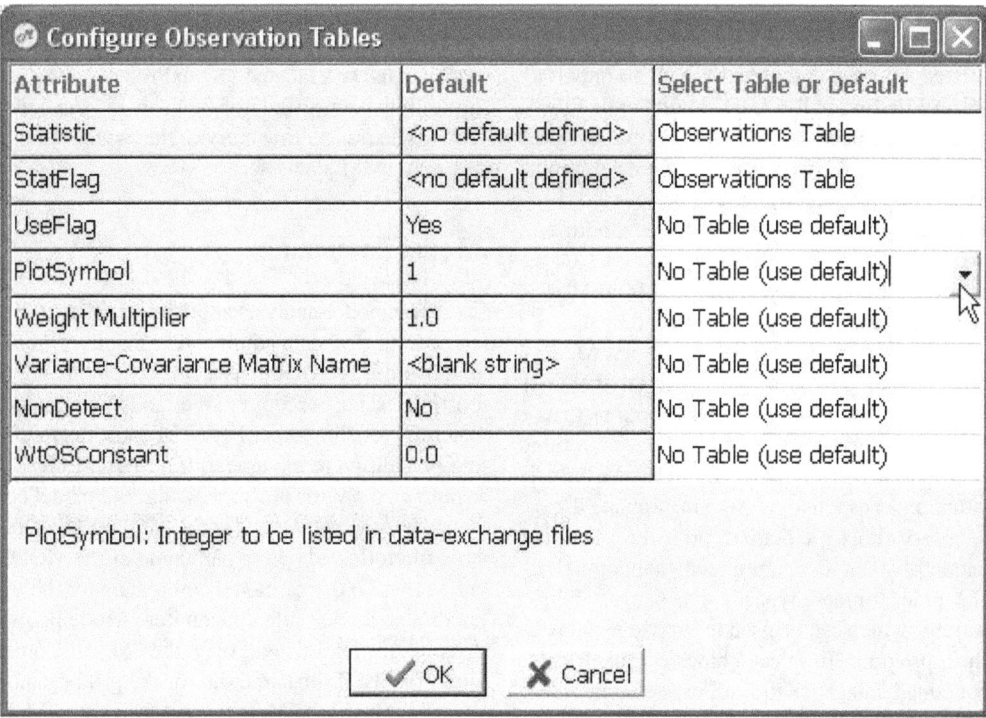

Figure 3. Configure Observation Tables window.

When the main UCODE input file is prepared, Model-Mate writes data according to each table configuration in effect at the time. Note that attributes that have been configured as "No Table (use default)" do not appear in either the member or group table and are not written to the UCODE input file. As a result, UCODE will use default values for these attributes. ModelMate saves any nondefault values assigned to attributes, and it displays them when configured to do so. However, values for attributes that do not appear in the member or group table are not exported to the UCODE input file.

The "Configure Parameters Table" window, accessed by clicking the "Configure Tables" button on the Parameters tab of the main ModelMate window (fig. 1), includes in the Attribute column an entry labeled "Derived." The Derived attribute is a special case. The Derived attribute can be shown in the Parameters table or hidden. Derived parameters are defined separately (see section titled "Derived Parameters"). Merely showing or hiding the Derived attribute in the Parameters table does not affect UCODE input or operation. If the Derived attribute is shown in the Parameters table, however, the checkboxes in the Derived column can be used to specify whether a parameter is defined as a derived parameter or not in UCODE input. If the checkbox is unchecked, the parameter is an ordinary parameter and the value listed in the "Start Value"

column is used. If the checkbox is checked, an expression that defines the value of the derived parameter as a function of one or more other parameters must be provided in the "Derived Parameters" window, and the value in the "Start Value" column of the Parameters table is ignored. See the section titled "Derived Parameters" for details.

Bringing Data into ModelMate

Before starting to use ModelMate on a modeling project, the user needs to have a functioning process model. Instructions for importing data for a MODFLOW-2005 model are provided in the section titled "Connection with MODFLOW-2005" under "Application/Model Connection." Data for MODFLOW-2005 models built with ModelMuse (Winston, 2009) can be exported by ModelMuse and read by ModelMate as described in the section titled "Use of Model-Mate with ModelMuse."

If input file(s) for UCODE have already been prepared using a text editor or other software before a ModelMate project has been started, the user can import the main UCODE file and the contents will be used to populate the ModelMate project. The model referenced in the UCODE file does not need to be a MODFLOW-2005 model. This method can be used for any UCODE file and any model that reads text file(s) as input and writes simulated values to output text file(s). To import a UCODE file, select **File|Import|UCODE Main Input File As New Project** from the menu bar (fig. 1), then browse to select the desired UCODE input file. By convention, UCODE input files have the extension "in," but any file can be selected. If the current project contains unsaved changes, the user is notified and prompted to save the current project. Before the UCODE input file is imported, all project data are cleared from ModelMate. Most or all data contained in or referenced by the UCODE input file are imported and stored in the new Model-Mate project. Data related to UCODE capabilities that are not supported by ModelMate are ignored. UCODE options that are not supported by ModelMate are listed in the section titled "Use of UCODE Options Not Supported By ModelMate."

In some situations a user may wish to manipulate data for parameters, observations, predictions, prior-information, or their associated groups in ways not directly supported by ModelMate. The tables for these types of data support copy and paste operations, which can be used to exchange data with a spreadsheet program. To select a block of cells for export of the contained data, click the cell in one corner of the block to be selected, then <Shift>-click the cell in the opposite corner of the block of cells. Use <Ctrl>-C to copy the data to the Windows clipboard. The data generally can be pasted into an external spreadsheet program using <Ctrl>-V or other command supported by the external program. Copying data back to the ModelMate table is the reverse operation. Use the external program to copy a block of data to the Windows clipboard. Then use <Ctrl>-V to paste the data into the appropriate ModelMate table; the data will be pasted into the table such that the upper left corner of the block of data will be pasted into the currently selected cell. If parameter names are to be included in the copied and pasted data, click the "Configure Tables" button on the Parameters tab to open the "Configure Parameter Tables" window and ensure that the "Freeze Parameter Names" box is unchecked.

Project Name

Each ModelMate project stores a project name, which by default in a new project is the base filename of the ModelMate project file. The project name is displayed in the left panel of the status bar at the bottom of the main ModelMate window. The project name is used in generating the name of a batch file that can be used at the operating system prompt to invoke UCODE, and it is used in generating the name of a template file for creating a MODFLOW-2005 Parameter Value (PVAL) file when importing parameters from MODFLOW-2005 (see section titled "Importing Parameter Information from MODFLOW-2005 Input Files"). By default, the project name is changed whenever the project file is saved with a new name, to keep the project name the same as the project file name. The project name can be changed by selecting menu item **Project|Project Name and Title**. When this menu item

is selected, a window titled "Project Name and Title" opens. To assign a project name different from the project file name, uncheck the box labeled "Keep Project Name Same As Project (.mtc) File Name" and assign a new project name where indicated. The project title entry in this window currently (2011) is not used by ModelMate.

Model Commands

The model-analysis application needs to be provided an operating-system command to invoke a model run; this command is referred to as a "model command." If pre- or postprocessing is required, the model command should be a batch file or similar script that invokes the required processing in addition to the model. If a MODFLOW-2005 model requires no pre- or postprocessing, the model command can consist of the pathname for the MODFLOW-2005 executable file followed by the pathname of the MODFLOW-2005 name file. To define model commands for the primary and(or) predictive model, select menu item **Model|Commands to Invoke Model**; this will open the "Model Commands" window (fig. 4). If the name file for the primary model has been stored in the ModelMate project (see section titled "Model Settings"), and if the model run does not require pre- or postprocessing, a model command can be generated by clicking the button labeled "Generate Command for Forward Primary MODFLOW Run." A model command for the predictive model similarly may be generated by clicking the "Generate Command for Forward Predictive MODFLOW Run" button. If a batch file or other script is to be used as a model command, use the fields provided to supply the file name, or click the file browser button (![icon]) to browse and select the batch or script file.

Application/Model Connection

UCODE interacts with models by manipulating model-input files, invoking the model, and extracting values from model-output files. ModelMate explicitly supports the model software MODFLOW-2005 (Harbaugh, 2005) in that it can read input files for a model based on MODFLOW-2005 and use the contents of those files to define parameters, observations, and predictions. As part of the import process, Model-Mate generates the template files and extraction-instruction files required by UCODE. Although other model software currently is not explicitly supported, users likely will find ModelMate to be a useful tool for preparing UCODE input, invoking UCODE, and interpreting UCODE results when other model software is used.

In the course of model analysis, UCODE prepares text files to be used as input by the model software. If the model software requires that model parameters be provided in files other than text files, a preprocessor utility capable of reading text files generated by UCODE and preparing input

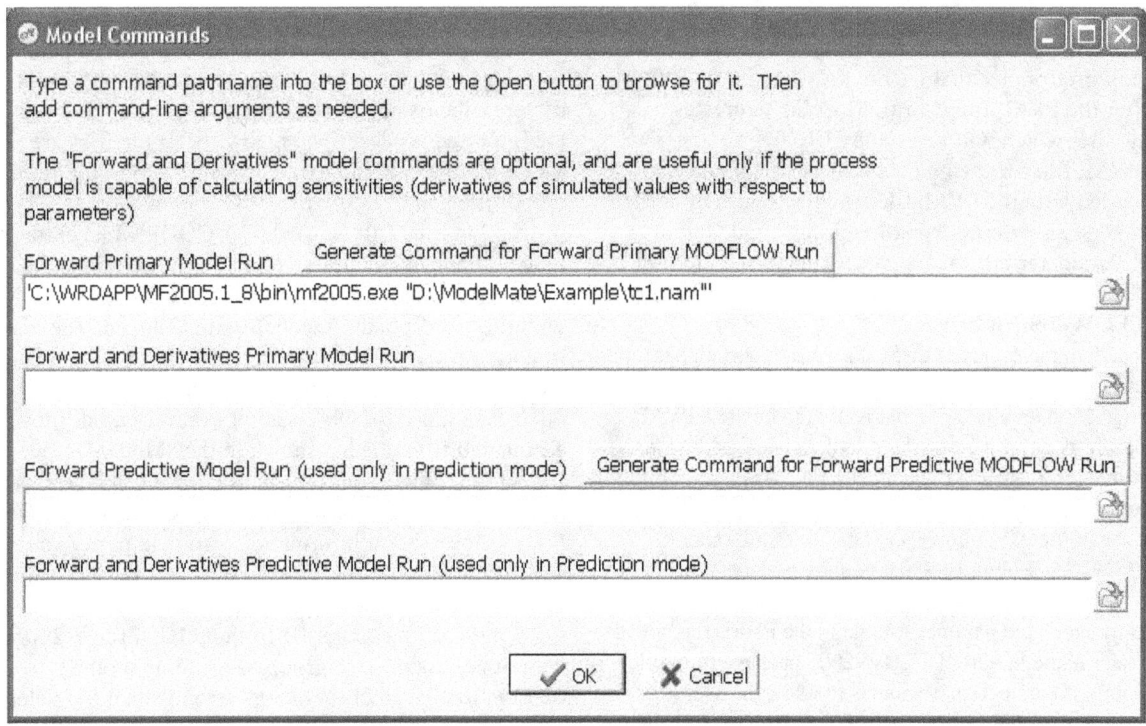

Figure 4. Model Commands window.

files readable by the model software can be used. Similarly, UCODE is designed to read model-simulated values from text model-output files. If the simulated values are in files other than text files, a postprocessor can be used to process the model output file(s) and prepare a text file containing the model-simulated values of interest.

When starting a ModelMate project, users have a number of options for populating the project with data. The following sections describe these options.

Connection with MODFLOW-2005

ModelMate supports two convenient ways for setting up the connection between UCODE and MODFLOW-2005. If software other than ModelMuse (Winston, 2009) is used to build the model, ModelMate can read the MODFLOW-2005 PVAL file (Harbaugh, 2005) to define parameters, and it can read MODFLOW-2005 Observation Process input files (Harbaugh, 2005) to define either observations or predictions. For a model built with ModelMuse, the user has the additional option to use ModelMuse to generate a partly populated ModelMate file containing basic information about parameters, observations, and predictions.

Importing Parameter Information from MODFLOW-2005 Input Files

Parameters (name and value) can be defined by importing data from MODFLOW-2005 input files listed in a name file. Currently (2011), only parameters listed in the PVAL file of a MODFLOW-2005 dataset are imported. To import parameters, select menu item **File|Import|Parameters from MODFLOW-2005**. Browse to select a MODFLOW-2005 name file that lists a PVAL file. When the name file is selected, it is opened and several pieces of information are obtained from it and stored in the ModelMate project. The relative directory path and file name of the name file are stored. If the ModelMate project already has parameters listed in the Parameters table, the user is given the choice of replacing the existing parameters or canceling the operation. Importing parameters from a MODFLOW-2005 dataset always deletes all existing parameters and adds all parameters listed in the PVAL file. If parameters that are not listed in the PVAL file are needed in addition to parameters to be imported, the additional parameters need to be added after the parameters listed in the PVAL file are imported.

Parameter names and values are read from the PVAL file and used to populate the Parameters table. Defaults are assigned for attributes that are not provided in the PVAL file. After the PVAL file is read, ModelMate creates a template file, which will be used by UCODE to generate a new PVAL file for each model run. The template file is associated with the PVAL file, and file names of both are stored (see section titled "Model Input and Output Files"). Parameter information also is imported in this way when **File|Import|Parameters and Observations from MODFLOW-2005** is executed.

Once parameter names and values are imported, parameter names can be changed. By default, if parameter names are changed, the template file that is used to create a PVAL file for each model run is rewritten with the new parameter names (to avoid rewriting the template file, as when multiple template files are used, see section titled "Model Settings"). For the renaming of parameters to be successful, other MODFLOW-2005 input files that contain parameter names need to be modified accordingly; otherwise, MODFLOW-2005 will fail to match the parameter name in the PVAL file with a parameter name in a MODFLOW-2005 package input file. If renaming of parameters cannot be avoided, the user may prefer to edit the MODFLOW-2005 input files first and then re-import parameters.

Importing Observation Information from MODFLOW-2005 Input Files

Observation names and observed values can be imported from MODFLOW-2005 Observation Process input files (Harbaugh and Hill, 2009). To import observations, select **File|Import|Observations from MODFLOW-2005** (from menu bar, fig. 1). Browse to select a MODFLOW-2005 name file that lists one or more Observation Process input files; this should be the same name file that was selected when importing parameters, if parameters have already been imported. When the name file is selected, it is opened and several pieces of information are obtained from it and stored in the ModelMate project. The relative directory path and file name of the name file are stored. If the ModelMate project already has observations listed in the Observations table of the Observations window, the user is given the choice of replacing the existing observations or canceling the operation. Importing observations from a MODFLOW-2005 dataset always deletes all existing observations and adds all observations in Observation Process input files referenced in the name file. If observations that are not listed in the Observation Process input files are needed in addition to observations to be imported, the additional observations need to be added after the observations listed in the Observation Process input file are imported. Observation names and observed values are read and used to populate the Observations table (fig. 5). Defaults are assigned for attributes (for example, group name, plot symbol, and others) that are

not provided in the Observation Process input file. Note that MODFLOW-2005 input does not include statistics and flags related to calculation of observation weights; these attributes need to be assigned after observations are imported, as described below. After the files are read, ModelMate creates an extraction-instruction file for each output file generated by the Observation Process. The instruction files will be used by UCODE when extracting model-simulated values from model-output files after each model run completes. Each instruction file is associated with the corresponding model-output file, and file names of both are stored (see section titled "Model Input and Output Files"). Pairs of model-output and instruction files other than those for Observation Process output files are eliminated from the list shown in the "Model Output and Instruction Files" window; these entries can be recreated as described in the section titled "Model Input and Output Files." Observation information also is imported in this way when **File|Import|Parameters and Observations from MODFLOW-2005** is executed.

Observations imported from a MODFLOW-2005 dataset are assigned to groups depending on the type of observation. Head observations (read from a file listed with type HOB in the name file) are assigned to group "Heads" for observations that are hydraulic heads (observations in the HOB file for which ITT equals 1) and to group "Head_Changes" for observations that are changes in head (observations for which ITT equals 2). Drain-boundary flow observations (file type DROB) are assigned to group "DRN_flows." River-boundary seepage observations (file type RVOB) are assigned to group "RIV_flows." General-head-boundary flow observations (file type GBOB) are assigned to group "GHB_flows." Constant-head-boundary flow observations (file type CHOB) are assigned to group "CHOB_flows." Group names can be edited after the import process is complete. A common approach is to assign Statistic and StatFlag to groups and to assign observations to observation groups accordingly. New groups can be added and used as deemed appropriate to facilitate assignment of these and other attributes (see section titled "Parameter, Observation, and Prediction Groups").

MODFLOW-2005 Observation Process input files do not contain information related to the weight to be assigned each observation. After observations have been imported, the "Observations" window (fig. 5) opens to remind and enable the user to assign Statistic and StatFlag values to the observations or observation groups. UCODE input instructions (Poeter and others, 2005, p. 83) state that Statistic is a value from which a weight is calculated for each observation and that StatFlag defines the Statistic and determines how it is used to calculate the weight. Statistic may be a variance (select "VAR"), standard deviation ("SD"), coefficient of variation ("CV"), weight ("WT"), or square root of the weight ("SQRWT"). When assigning Statistic values, it is essential to ensure that StatFlag is assigned appropriately for each Statistic.

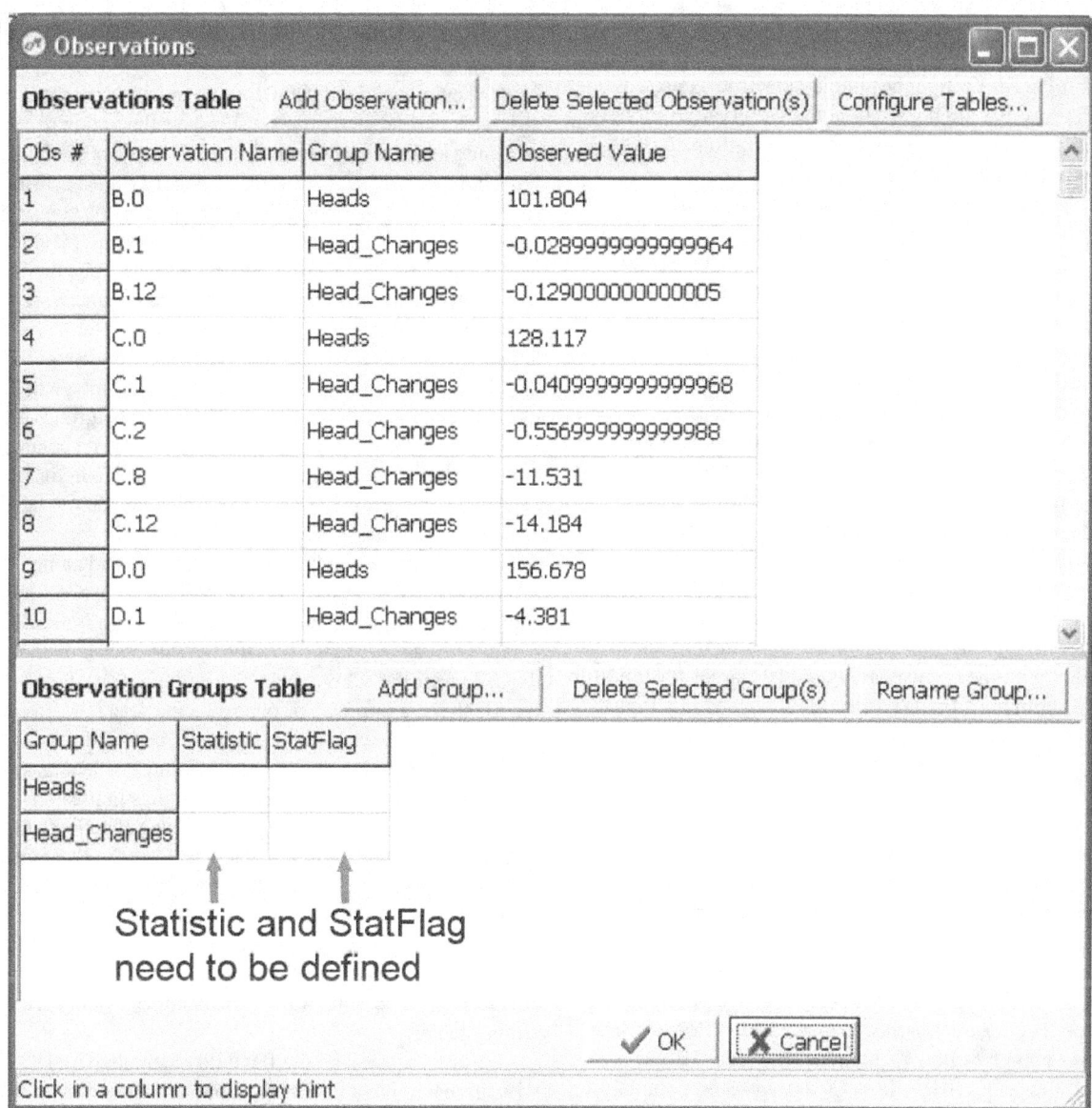

Figure 5.   Observations window after importing observations from MODFLOW-2005.

Importing Prediction Information from MODFLOW-2005 Input Files

When a MODFLOW-2005 model is used as a predictive model, it is convenient to use the MODFLOW-2005 Observation Process to calculate and write model-simulated values to be treated as predictions. When a model is set up this way, ModelMate can import information in the Observation Process input files and store it as information related to predictions.

To import Observation Process information into ModelMate as predictions, use **File|Import|MODFLOW-2005 Observations As Predictions** (from menu bar, fig. 1).

Browse to select a MODFLOW-2005 name file that lists one or more Observation Process input files; this would be a name file for a predictive model. When a name file is selected, it is opened and several pieces of information are obtained from it and stored in the ModelMate project. The relative directory path and file name of the name file for the predictive model are stored. If the ModelMate project already has predictions listed in the Predictions table of the Predictions window, the user is given the choice of replacing the existing predictions or canceling the operation. The Predictions window is similar in appearance and functionality to the Observations window (fig. 5). Importing predictions from a MODFLOW-2005 dataset always deletes all existing predictions and adds all observations in the Observation Process input files referenced in the name file as predictions. If predictions that are not listed in the Observation Process input files are needed in addition to observations to be imported, the additional predictions need to be added after the predictions listed in the Observation Process input file are imported. On import, observation names (MODFLOW variable OBSNAM) are stored as prediction names, and observed values (MODFLOW variable HOBS) are stored as reference values. Defaults are assigned for attributes (for example, group name, plot symbol, and others) that are not provided in the Observation Process file. Note that MODFLOW-2005 input does not include statistics and flags related to calculation of weights; these attributes (MeasStatistic and MeasStatFlag) need to be assigned after observations are imported as predictions, as described below. After the files are read, ModelMate creates an extraction-instruction file for each output file generated by the Observation Process. The instruction files will be used by UCODE when extracting model-simulated values from model-output files after each model run completes. Each instruction file is associated with the corresponding model-output file, and file names of both are stored (see section titled "Model Input and Output Files").

Predictions imported from Observation Process input files are assigned to groups depending on the type of observation. Head predictions (read from a file listed with type HOB in the name file) are assigned to group "Heads" for predictions that are hydraulic heads (entries in the HOB file for which ITT equals 1) and to group "Head_Changes" for predictions that are changes in head (ITT equals 2). Drain-boundary flow predictions (file type DROB) are assigned to group "DRN_flows." River-boundary seepage predictions (file type RVOB) are assigned to group "RIV_flows." General-head-boundary flow predictions (file type GBOB) are assigned to group "GHB_flows." Constant-head-boundary flow predictions (file type CHOB) are assigned to group "CHOB_flows." Group names can be edited after the import process is complete. New groups can be added and used as deemed appropriate (see section titled "Parameter, Observation, and Prediction Groups").

MODFLOW-2005 Observation Process input files do not contain information that can be related to expected prediction variances. After predictions have been imported, the "Predictions" window opens to remind and enable the user to assign MeasStatistic and MeasStatFlag values to the predictions or prediction groups. UCODE input instructions (Poeter and others, 2005, p. 90) state that MeasStatistic is a value from which a variance is calculated for each prediction, and that MeasStatFlag defines the MeasStatistic and determines how it is used to calculate the variance. MeasStatistic can be either a variance (select "VAR") or a standard deviation ("SD"). When assigning MeasStatistic values, it is essential to ensure that MeasStatFlag is assigned appropriately for each MeasStatistic.

Parameter, Observation, and Prediction Groups

When organizing UCODE input to define parameters, observations, and predictions, users have the option of assigning attributes by group or to each member of a group. For example, a subset of parameters can be log-transformed for parameter estimation. UCODE users have the choice of defining parameter groups and specifying that all parameters in a group are to be log-transformed. Alternatively, the log-transform option can be specified for each parameter, regardless of its membership in a group. Each parameter belongs to exactly one group; by default, parameters are assigned to the ParamDefault group. However, to take advantage of the UCODE capability to report analysis results by group or to more conveniently organize the application/model connection, users generally define their own groups. To define a new parameter group in ModelMate, click the button labeled "Add Group" in the Parameters tab of the main window. In the "Add Parameter Group" window, provide a valid name (1 to 12 characters, first character is a letter) and click OK. After adding groups as desired, the group designation for individual parameters can be changed. In the Parameters table, click the cell under the "Group Name" heading for the parameter for which the group is to be changed, and select the group to which the parameter is to be assigned. Then, attributes that appear as headings in the Parameter Groups table are assigned by group. Attributes that appear as headings in the Parameters tables are assigned by parameter. Attributes that do not appear as headings in either the Parameters table or the Parameter Groups table (fig. 1) are not included in UCODE input and thus are assigned default values by UCODE.

The windows used to define Observations and Predictions also contain tables of members and groups. These tables operate essentially the same way as the Parameters and Parameter Groups tables.

In accordance with the flexibility supported by UCODE with respect to input of data to define members and groups, the tables used to define parameters, observations, and predictions are user-configurable. For information on configuring members and groups tables, see the section titled "Table Configuration for Members and Groups."

Model Input and Output Files

UCODE interacts with primary and predictive models by preparing model-input files containing numeric values that depend on parameter values, executing the model, and extracting model-simulated values from model-output files. Each model-input file has a corresponding template file, which is used to guide construction of the model-input file. Similarly, each model-output file has a corresponding instruction file, which contains instructions that enable UCODE to extract model-simulated values from the model-output file and associate each simulated value with a specified observation or prediction.

For the primary model, the names of the model-input and template files can be viewed or set in the "Model-Input and Template Files" window (fig. 6). This window is accessed by selecting menu item **Model|Model-Input and Template Files**. Each row in the table in this window holds the names of a model-input file and a corresponding template file. When a cell in the table is clicked, a button appears that can be used to browse and select a file name. To add a row to the table, click on the lowermost visible cell and press the down arrow key on the keyboard. To delete a row, click in a cell of a row to be deleted and press the key combination <Ctrl>-Delete on the keyboard. To insert a row, click in a cell in a row and press the key combination <Ctrl>-Insert; a row will be inserted above the selected row.

The model-output and instruction file names are viewed or set in the "Model-Output and Instruction Files" window (fig. 7). To access this window, select **Model|Model-Output and Instruction Files**. Each row in the table in this window holds the names of a model-output file and a corresponding extraction-instruction file. When a cell in the table is clicked, a button appears that can be used to browse and select a file name. The processes for adding, deleting, and inserting rows are as described for the "Model-Input and Template Files" window.

For the predictive model, the windows that allow the user to view and set model-input, template, model-output, and instruction file names are like those for the primary model, except the window titles indicate the files are for the predictive model. To access these windows, select either

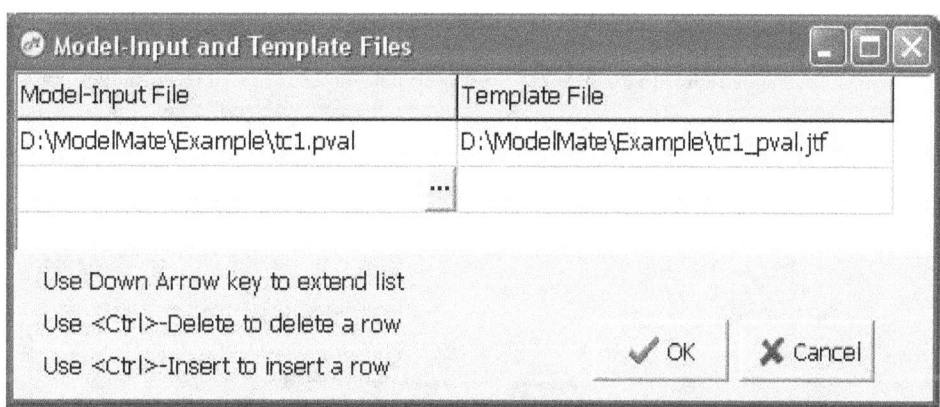

Figure 6. Model-Input and Template Files window.

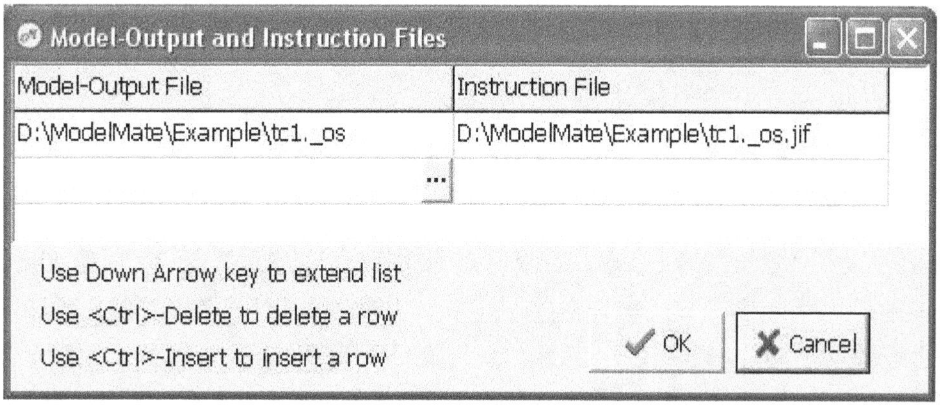

Figure 7. Model-Output and Instruction Files window.

Model|Predictive-Model Input and Template Files or **Model|Predictive-Model Output and Instruction Files**.

When the user invokes one of the **File|Import** commands to read a MODFLOW-2005 dataset and import parameters, observations, or predictions, ModelMate generates either a template file (when parameters are being imported) or an instruction file (when observations or predictions are being imported). In some situations, template and instruction files cannot be generated automatically. Template files cannot be generated automatically for a model based on model software other than MODFLOW-2005 or for a MODFLOW-2005 input file other than the PVAL file. Instruction files cannot be generated automatically for model software other than MODFLOW-2005 or for MODFLOW-2005 output files other than those referenced in the MODFLOW-2005 Observation Process input files. However, user-prepared template and instruction files can be used by ModelMate. Instructions are provided in Poeter and others (2005) for preparing template files (p. 109–110) and instruction files (p. 113–129). For additional details related to construction of template and instruction files, please see Banta and others (2006, Appendix A). When user-prepared template files are used, be sure to uncheck the box labeled "Link Template File with Parameters Table" in the Model Settings window (see next section).

Model Settings

When ModelMate imports parameter, observation, or prediction data from a MODFLOW-2005 model, the name file for the model is stored. As long as the name files for the primary and predictive model remain the same, the file names stored by ModelMate can be left as assigned. Similarly, parameter names read from a MODFLOW-2005 PVAL file are stored and generally can be left as assigned. In some situations, the user may want to use a different name file for either the primary or predictive model. To change the file names that are stored for these files, select menu item **Model|Model Settings**; this will open the "Model Settings" window (fig. 8). When the name of a name file is changed in the Model Settings window, the new file name is used when, in the "Model Commands" window, the user subsequently clicks one of the "Generate Command…" buttons (see section titled "Model Commands").

In the "Model Settings" window (fig. 8) is a checkbox labeled "Link Template File with Parameters Table." This checkbox is used to determine if the template file for the PVAL file should be rewritten if the number or names of parameters listed in the Parameters table of the ModelMate main window have changed. If it is checked and one or more parameters are deleted or added, or if a parameter name has been changed, a new PVAL template file is

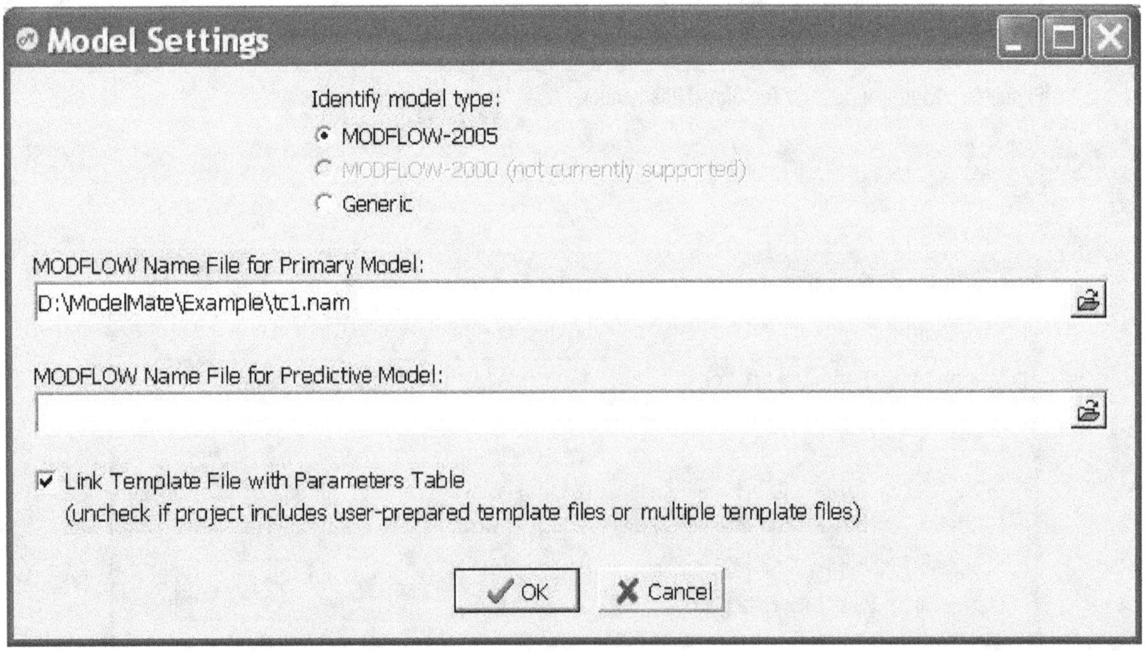

Figure 8.   Model Settings window.

written when the UCODE input file is recreated. In some cases, more than one template file will be used in order for UCODE to generate more than one model-input file. In other cases, the template file may be user-prepared. In such cases, clear the checkbox to avoid overwriting a valid template file with an invalid file.

Model Analysis With UCODE

Model analysis, in the context of this documentation, refers to analyses that can be performed to examine and quantify relations among observations, simulated values, residuals, parameters, predictions, uncertainty, and model linearity. The model analyses performed by UCODE and supported by ModelMate include the following UCODE modes: forward (to calculate residuals and an objective function), sensitivity analysis, parameter estimation, and prediction. The following model-analysis programs, which are included in the UCODE distribution and documented by Poeter and others (2005), also are supported by ModelMate: Residual_Analysis (for analysis of residuals) and Residual_Analysis_Adv (advanced residual analysis). This report does not explain the theory or use of the various capabilities of UCODE; for documentation of those capabilities, please refer to Poeter and others (2005) or Hill and Tiedeman (2007). Parameter estimation in particular is a process that generally is not amenable to a "cookbook" approach. Such issues as model parameterization, weighting of observations and prior information, model linearity,

parameter-estimation numerical methods, parameter correlation, parameter uncertainty, log-transformation of parameters, interpretation of statistics related to parameters and observations, and interpretation of graphs produced by GW_Chart (Winston, 2000) from UCODE output commonly need to be considered in the course of calibrating a model. These topics are beyond the scope of this report. Additional in-depth information on these and other topics related to parameter estimation, sensitivity analysis, and other model analyses can be found in Hill and Tiedeman (2007). This report primarily provides guidance on running ModelMate as a tool for preparing input for UCODE and related model-analysis and graphing applications.

Supported UCODE Modes

UCODE can be run in any of eight modes (Poeter and others, 2005, p. 31). Table 2 lists the modes and provides summaries for the modes that are supported in Model-Mate. The mode is selected on the left side of the main ModelMate window (fig. 1) when the selected model-analysis application is UCODE (the only choice in the initial version of ModelMate). When either the "Create UCODE Input Files" or "Run UCODE" button is clicked, ModelMate writes a UCODE input file constructed to run UCODE in the selected mode, and template and extraction-instruction files specific to MODFLOW-2005 are rewritten if required. The use of ModelMate to define details of UCODE operation in the various modes is described in the section titled "Setting UCODE Options."

Table 2. UCODE modes.

UCODE Mode	Summary
Forward	Primary model is executed once.
Sensitivity Analysis	For perturbation sensitivities, the primary model is executed once with unperturbed parameters plus once or twice for each adjustable parameter, depending on Sensitivity Method.
Parameter Estimation	Primary model is executed as for Sensitivity-Analysis mode for each parameter-estimation iteration plus one (the last iteration is for final statistics). Parameter-estimation iterations are executed until convergence criteria are met or the limit on number of iterations is reached.
Test Model Linearity	Not supported in ModelMate.
Prediction	Either Forward or Sensitivity-Analysis mode is run using predictive model.
Advanced Test Model Linearity	Not supported in ModelMate.
Nonlinear Uncertainty	Not supported in ModelMate.
Investigate Objective Function	Not supported in ModelMate.

Setting UCODE Options

Controls related to the analysis capabilities of UCODE are accessed by way of the **UCODE** menu. Each item on the **UCODE** menu opens a window that allows the user to control particular details of UCODE operation, as summarized in table 3. Each item is described in following sections.

Table 3. UCODE menu items.

UCODE menu item	Summary of controls
File Names	Names for UCODE main input file for primary and predictive models; output prefix for files generated by UCODE for analysis of primary and predictive models.
Settings	Model name and measurement units, calculation of statistics, output options.
Parameter-Estimation Settings	Limits, tolerances, and other controls related to parameter estimation; choice of algorithm; model-generated values that should be ignored as simulated values; output options.
Derived Parameters	Definition and use of equations defining parameters as functions of other parameters.
Prior Information	Definition and use of linear prior-information equations and prior-information groups.

UCODE File Names

Selecting menu item **UCODE|File Names** (appendix) opens the "UCODE Input and Output File Names" window (fig. 9) containing an upper panel, which relates to the primary model, and a lower panel, which relates to the predictive model. In each panel, the user can specify a name for a main UCODE input file and an output prefix to be used by UCODE in naming of output files (Poeter and others, 2005, p. 29). The edit controls for the UCODE input files include a button that allows the user to browse the file system and select an appropriate file. The file name is displayed as an absolute pathname, but it is stored as a relative pathname.

UCODE Settings

Selecting **UCODE|Settings** opens the "UCODE Settings" window (fig. 10). This window allows the user to define a model name; length, time, and mass units for the model; value of standard error to be used in calculation of statistics; and various output options. In the initial version of Model-Mate, several controls in this window are disabled because the indicated functionality is not yet supported; these controls appear grey.

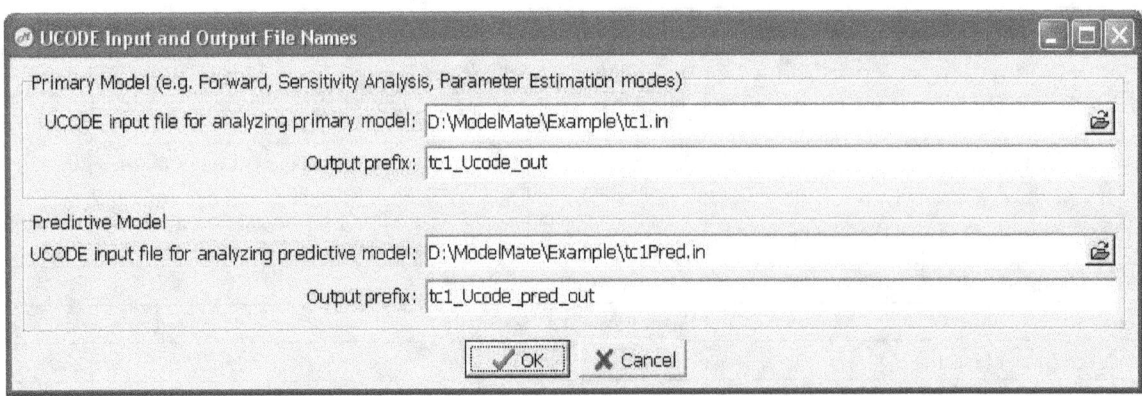

Figure 9.   UCODE Input and Output File Names window.

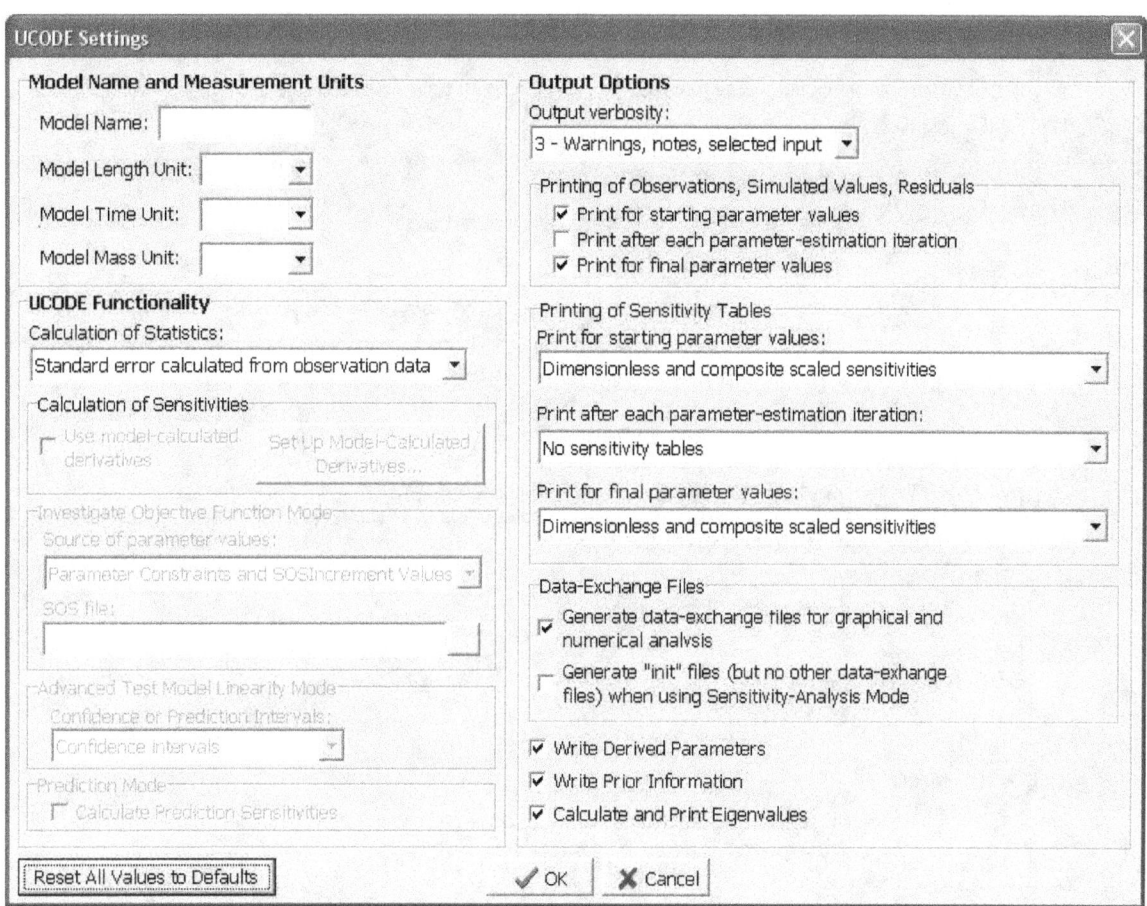

Figure 10.   UCODE Settings window.

Parameter-Estimation Settings

Selecting **UCODE|Parameter-Estimation Settings** opens the "UCODE Parameter-Estimation Settings" window (fig. 11). This window and the "UCODE Parameter-Estimation Advanced Settings" window (fig. 12), which is accessed by clicking the "Advanced Settings" button, allow setting values for variables of the Reg_GN_Controls input block (Poeter and others, 2005, p. 60–64), which in turn controls the parameter-estimation capabilities of UCODE. A short definition for each variable is displayed after the variable name. Poeter and others (2005) provide a detailed explanation of each variable. The Hookstep option of the Trust Region capability is described in the February 10, 2008, revision of Poeter and other (2005).

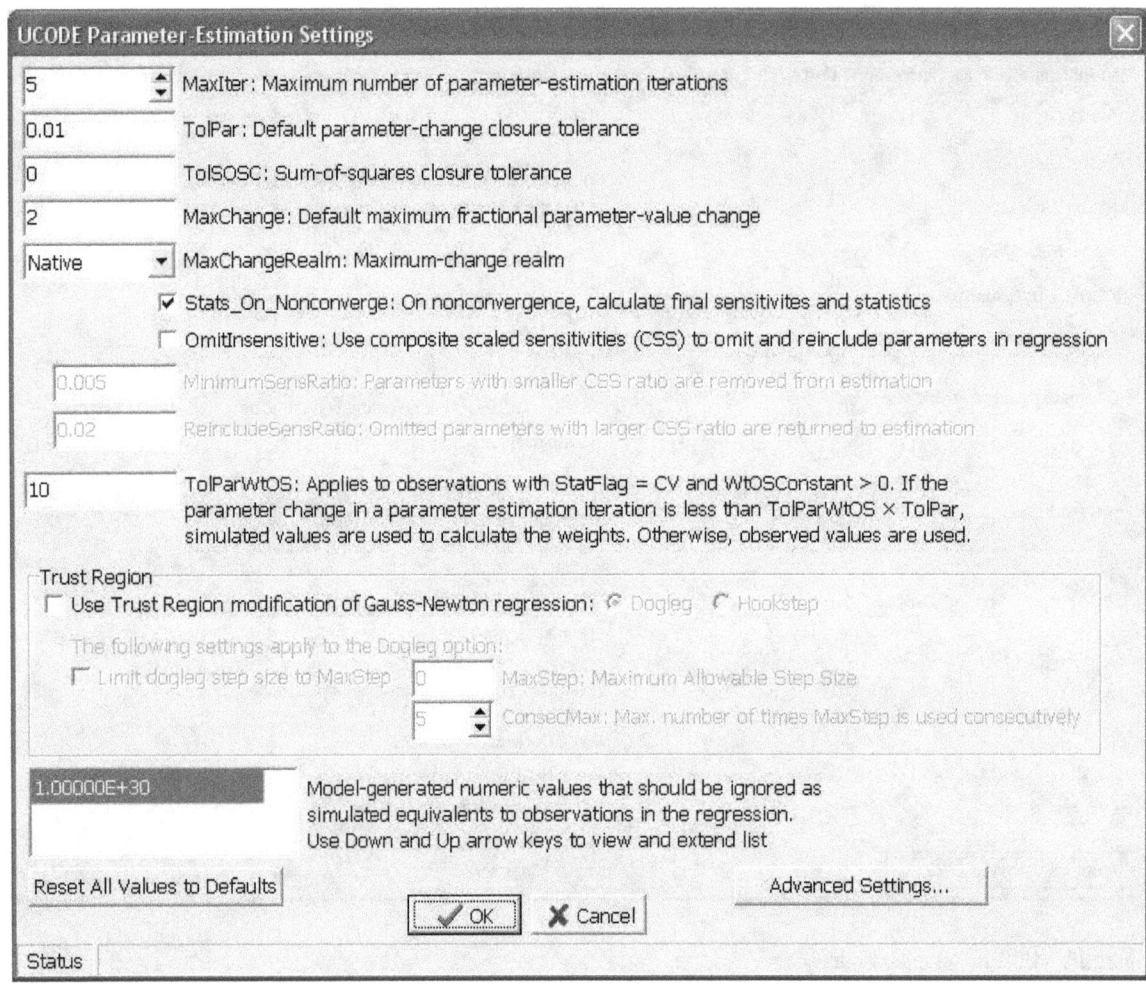

Figure 11. UCODE Parameter-Estimation Settings window.

Derived Parameters

Selecting **UCODE|Derived Parameters** opens the "UCODE Derived Parameters" window (fig. 13). Each parameter listed in the Parameters table on the Parameters tab on the right side of the main ModelMate window is listed in the "Derived Parameters" window. To define a parameter as a derived parameter, the Derived checkbox must be checked, and a valid equation must be entered in the Equation column. As described by Poeter and others (2005, p. 76–78 and 141–142), the "equation" is actually an expression; that is, just the equation right-hand side by which the derived parameter is calculated.

Once a valid equation is defined for a parameter, the Derived checkbox can be used to toggle the parameter's status between an ordinary parameter (unchecked) and a derived parameter. The equation is saved when the checkbox is in either state, but the equation is used and the parameter is derived only when the checkbox is checked. The Derived checkbox also can be shown on the Parameters table of the main ModelMate window (see section titled "Table Configuration for Members and Groups"). Checking or unchecking the Derived checkbox on the Parameters table has the same effect as in the Derived Parameters window; the checkbox status will be changed in both windows whenever one is checked or unchecked in either window.

Prior Information

Selecting **UCODE|Prior Information** opens the "Prior-Information Control" window (fig. 14). To define linear prior-information items, click the "Define Linear Prior Information"

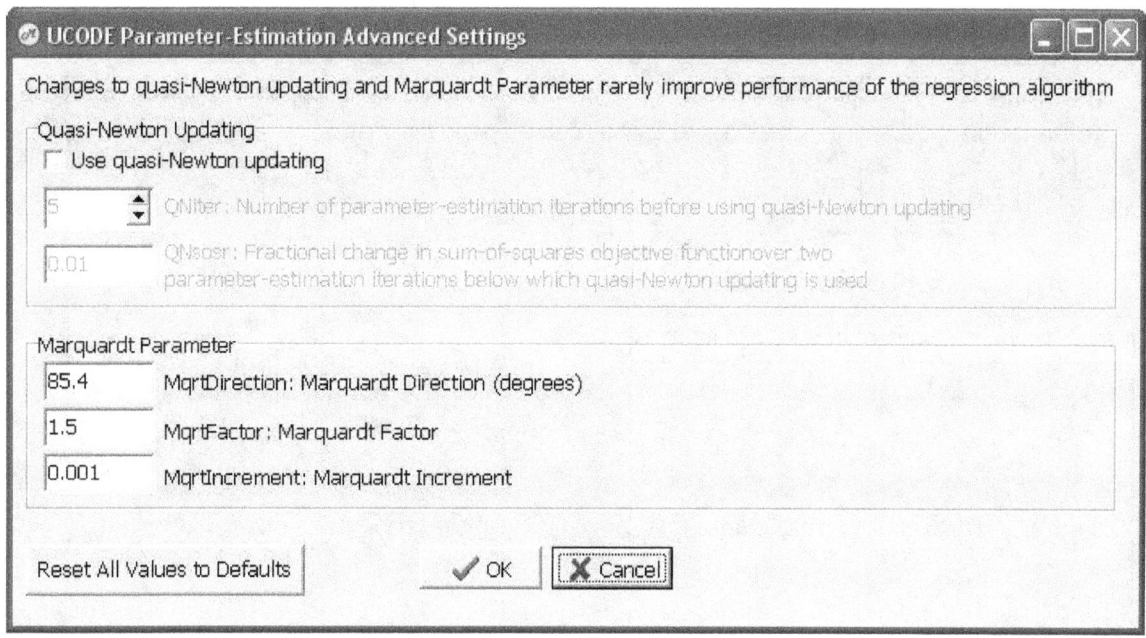

Figure 12.   UCODE Parameter-Estimation Advanced Settings window.

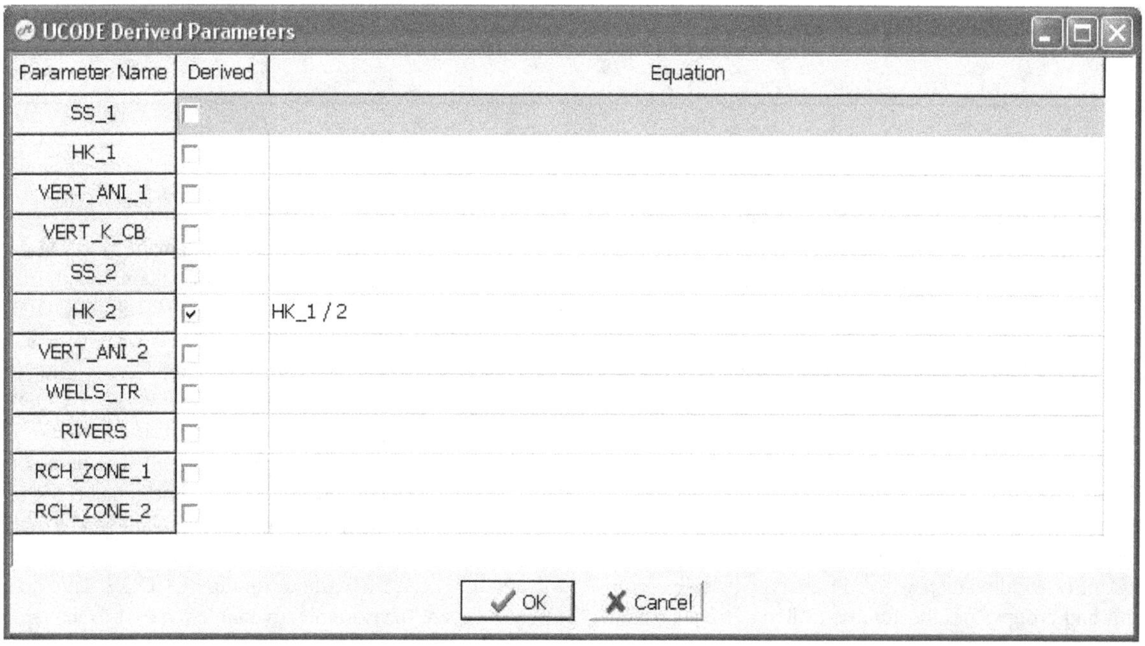

Figure 13.   UCODE Derived Parameters window.

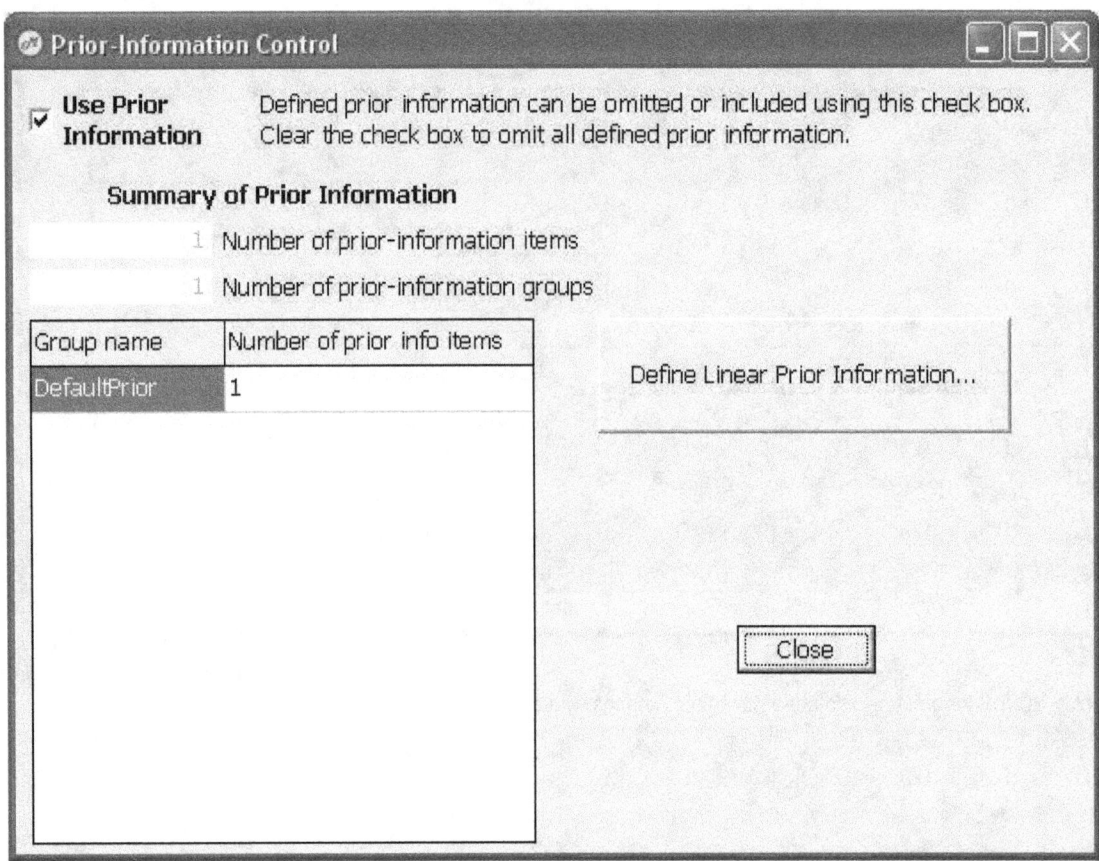

Figure 14.   Prior-Information Control window.

button. This will open the "UCODE Linear Prior Information" window (fig. 15), which contains a Prior-Information Table and a Prior-Information Groups Table. These tables can be configured to conform to the needs of the user (see section titled "Table Configuration for Members and Groups"). For each prior-information item desired, the user enters a name for the prior-information item, linear equation (again, actually an expression), prior-information value, and other pertinent data. To keep changes and close the "UCODE Linear Prior Information" window, click OK.

In the Prior-Information Control window is a checkbox labeled "Use Prior Information," which can be used to turn on or off the use of all prior information. Boxes in the Prior-Information Control window display the numbers of prior-information items and groups, and the number of items in each group. These boxes are not editable. The "Use Prior Information" setting is saved whenever the "Prior-Information Control" window is closed, which is done by clicking the Close button.

Running UCODE

Two buttons on the left side of the main ModelMate window (fig. 1) control the creation of a main UCODE input file and execution of UCODE. When the "Run UCODE" button is clicked, a UCODE main input file is written with the file name specified in the "UCODE Input and Output File Names" window, a batch file that can be used to invoke a UCODE run from the operating system is written, and (if creation of the input file is successful) UCODE is invoked. The name of the batch file is formed as "RunUcode_" concatenated with the project name (see section titled "Project Name and Title"). In some situations it may be desirable to create the main input file and the batch file without running UCODE. This option might be used, for example, to enable the user to manually edit the UCODE input file(s) to specify UCODE options not supported by ModelMate or to review the input file(s) before running UCODE. For situations like this, click the "Create

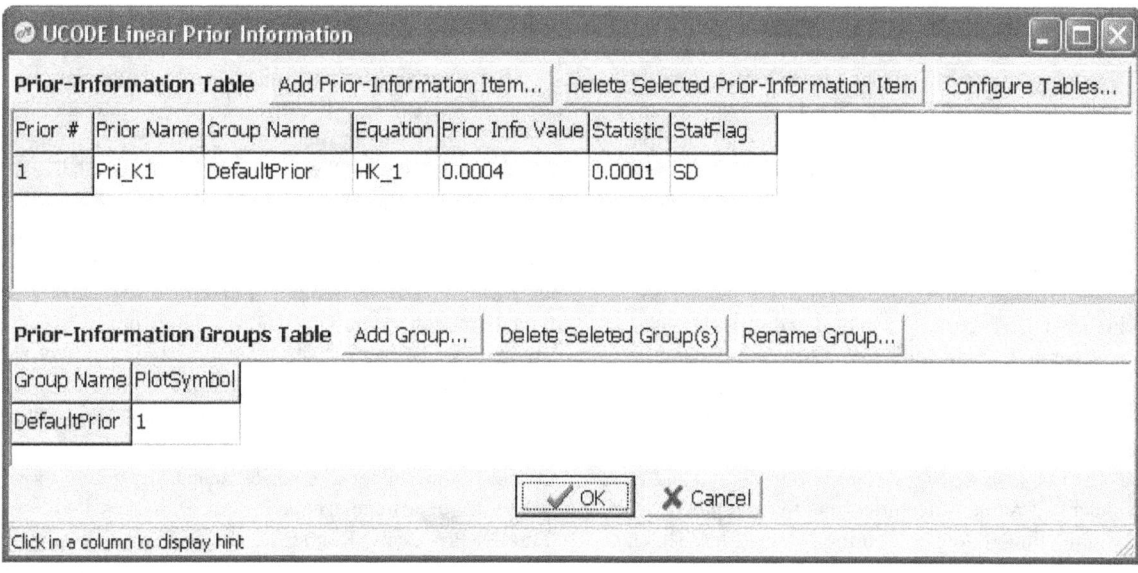

Figure 15. UCODE Linear Prior Information window.

UCODE Input Files" button. The main UCODE input file can then be viewed with the file viewer (see section titled "File Viewer"). If options not supported by ModelMate are needed, the input file(s) can be edited as needed (see Poeter and others, 2005, for instructions for preparing UCODE input), and the batch file can be executed at the operating-system prompt.

Commonly when using Parameter-Estimation mode of UCODE, a user will want to use parameter values estimated in a previous parameter-estimation run as starting parameter values in a subsequent UCODE run. UCODE writes optimal parameters at the conclusion of each run in parameter-estimation mode to a file with extension "_paopt." To import optimal parameter values from an _paopt file and use them to replace current parameter values, select **File|Import|Optimized Parameters (_paopt file)**. By default, ModelMate will look for a file named by concatenating the UCODE output prefix (see section titled "UCODE File Names") and "._paopt." A dialog box labeled "Confirm" will open asking if this is the file to be imported. To import this file, click "Yes." To choose a different file, click "No." Choosing "No" will open a browser window that can be used to open any file with extension "_paopt"; however, if the number of parameters or parameter names in the file do not match those currently in the ModelMate project, the import will not succeed.

Parallel Processing

Some model analyses can require numerous process-model runs, and in many cases large numbers of these runs are independent of each other. In calculating sensitivities by perturbation, for example, for each parameter-estimation iteration the number of runs required to populate the Jacobian (or sensitivity) matrix can be any value from N to 2*N, where N is the number of adjustable parameters, plus one more process-model run with unperturbed parameters to calculate residuals. The actual number of runs will depend on the sensitivity-calculation method, which can be specified by parameter or by parameter group. Sensitivities can be calculated by forward differences (which requires one perturbation run for each adjustable parameter) or by central differences (which requires two perturbation runs for each adjustable parameter). Execution time for a UCODE run in parameter-estimation or sensitivity-analysis mode can be substantially shortened by using parallel processing.

UCODE implements parallel processing using the dispatcher/runner protocol described in Poeter and others (2005, chapter 12, p. 131–139). In this protocol, the UCODE executable file is the dispatcher program and JRUNNER (Poeter and others, 2005, p. 133) is the runner program. Multiple "runner" directories are set up, and model-input and other required files are set up in each runner directory. Each runner directory is set up such that a model run can be made in each runner directory and can generate model-output files in its own directory. Input files in common directories can be shared by multiple runner directories, but output files need to be unique to each directory. However, for ease of maintenance of the runner directories, users likely will find it convenient to design model input and output such that all model input and output files are located in the specified runner directory or in a subdirectory located in the runner directory. The runner directories also should be separate from the directory where the dispatcher program makes model runs. Runner directories can be located on the same computer as the one where the dispatcher program is run, or they can be on other computers accessible by a network. The dispatcher program needs to have read and write permissions in all runner directories, and the computers need to use the same line-ending convention for text files. This last requirement means that when ModelMate is used to invoke UCODE, all computers used for parallel processing need to run the Windows operating system, because ModelMate runs only under Windows and UCODE input files prepared by ModelMate use the Windows line-ending convention.

Maintenance of Runner Directories

In setting up runner directories, it is important that all model-input files and files required by JRUNNER and UCODE to create model-input files are identical. The task of ensuring that all files are identical before each parallel-processing run can be cumbersome if handled manually. ModelMate provides a capability to update specified files in runner directories to match files in the directory where the dispatcher program makes model runs. The update procedure can

be run automatically before each invocation of UCODE, or it can be initiated interactively from the ModelMate GUI.

Before setting up the runner-directory maintenance capability in ModelMate, it is convenient to design model input so that all model-input files are located in the directory where the model is to be invoked or in subdirectories relative to that directory. In this discussion, the term "master directory" is used to refer to the directory where UCODE will invoke the process model. Runner directories can be on the same computer where UCODE will run or on other computers accessible to the computer where UCODE will run. Runner directories can be created either by using the operating system or from the "Parallel-Processing Runner Directories" window (fig. 16). It is not necessary to populate the runner directories with model-input files at this point.

To define runner directories, select menu item **Parallel Processing|Runner Directories** (appendix); this will open the "Parallel-Processing Runner Directories" window (fig. 16). Use each row in the table in this window to define one runner directory. Click on the lowermost visible row and press the down arrow key on your keyboard as needed to extend the list of runner directories. To delete a row, select the row containing the runner to be deleted and click the "Delete Selected Runner" button. In the Runner Directory column, click on the button with an image of an ellipsis ("…"), which will enable you to browse the file system using a dialog window labeled "Browse For Folder." Select an existing folder to be used as a runner directory and click OK; alternatively a new folder can be created and named using the "Make New Folder" button in the "Browse For Folder" window. The name of the directory will be used to populate the Runner Name column if that column is empty. Set the expected run time for each runner directory and click OK to save the runner-directory information and close the "Parallel-Processing Runner Directories" window.

Runner Name	Use?	Runner Directory	Expected Run Time (sec)
Runner1	☑	D:\ModelMate\Example\Runner1	1
Runner2	☑	D:\ModelMate\Example\Runner2	1
	☑		1

Use Down Arrow key to extend list

Delete Selected Runner   Sort By Expected Run Time   Move Runner Up   Move Runner Down

✓ OK   ✗ Cancel

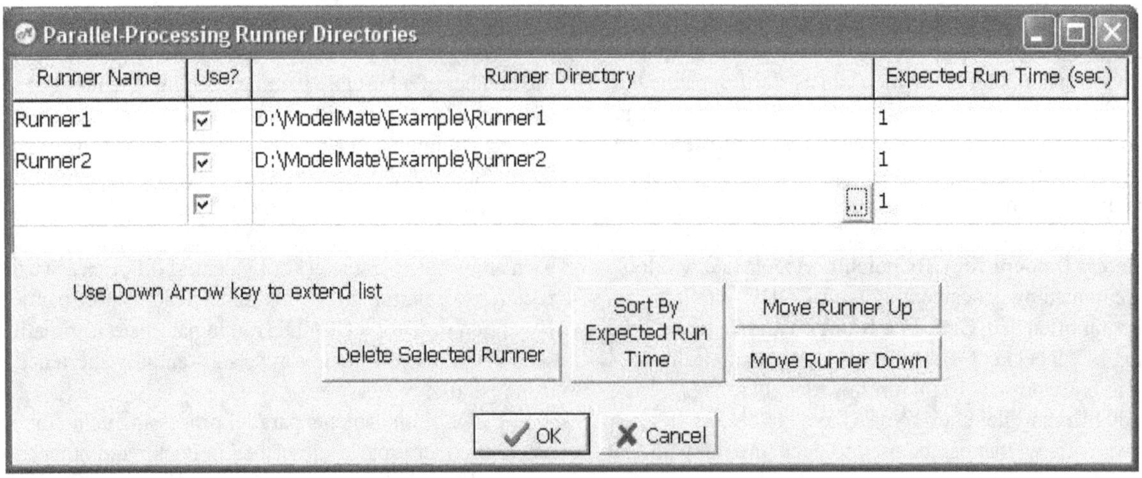

Figure 16. Parallel-Processing Runner Directories window.

The next step is to define a set of files that need to be identical in all runner directories. For most convenient maintenance, arrange model input in the master directory such that all input files are referenced by relative pathnames in a way that also will work in all runner directories. For example, if the process model is set up to read all input from files in a subdirectory named Model_Input, create a subdirectory named Model_Input under each of the runner directories as well. As the modeling effort proceeds, for changes to the model other than parameter adjustments (for example, conceptual model changes), ensure that changes initially are made to files in the master directory (or in subdirectories under the master directory). As these kinds of changes are made, the changes need to be propagated to the model-input files in the runner directories. Two categories of files need to be maintained to be identical between the master directory and the runner directories: (1) files that are directly prepared by the user, and (2) files that are used by UCODE (or

JRUNNER) to generate model-input files. Files whose contents do not depend on values of adjustable parameters are in the first category, and template files are in the second category. Model-input files that are to be generated by UCODE (or JRUNNER) using template files do not need to be maintained by ModelMate because these model-input files always will be generated by UCODE or JRUNNER before each model run. Users may want to include the jrunner.exe executable file as a maintained file, to ensure that it is accessible in each runner directory. Instruction files do not need to be copied to runner directories or maintained because extraction instructions are passed to the runner directories in files generated by UCODE.

To select files to be maintained as identical among the master and all runner directories, select **Parallel Processing|Parallel Control**; this will open the "UCODE Parallel-Processing Control" window (fig. 17). In this window, click the button labeled "Files for Runner Directories" to open

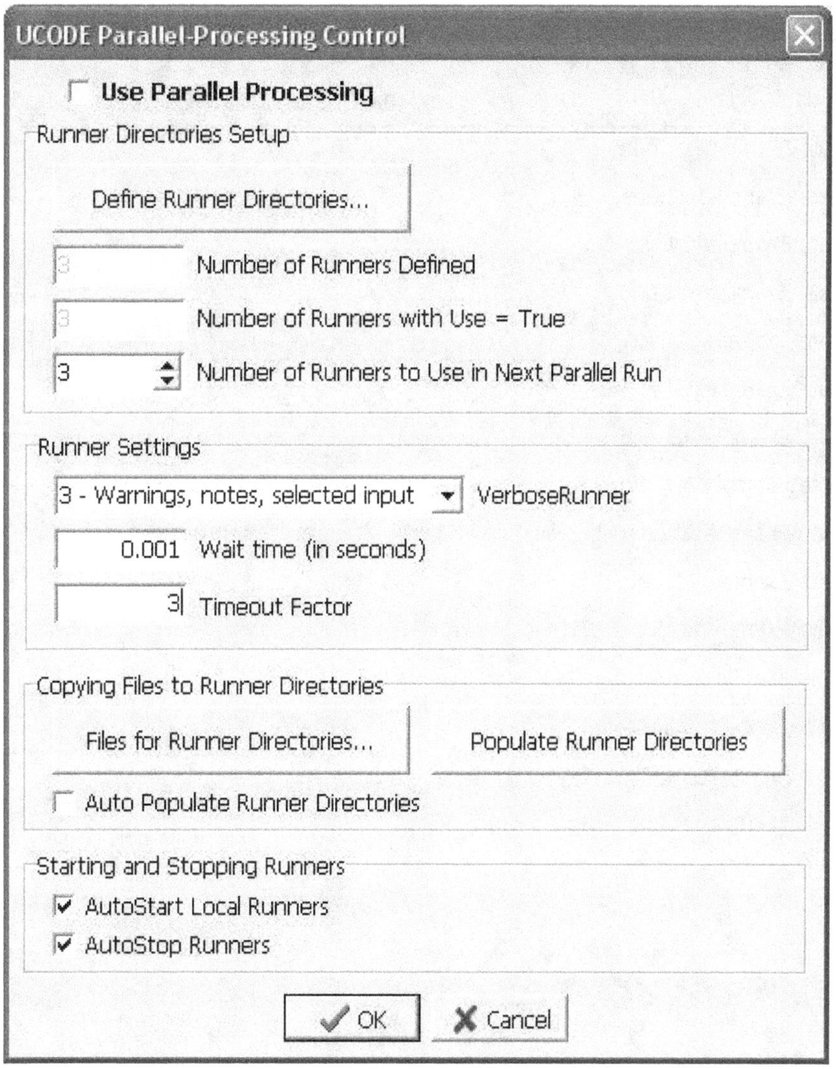

Figure 17.   UCODE Parallel-Processing Control window.

the window labeled "Files to be Copied to Runner Directories" (fig. 18). Use each row in the table in this window to define the name of a file in the master directory (or in a subdirectory under the master directory) to be maintained in the runner directories. Select all model-input files that are user-generated and all template files. Click OK to save the list of files and close the "Files to be Copied to Runner Directories" window. In the "UCODE Parallel-Processing Control" window, the runner directories can be updated at this or any time by clicking the button labeled "Populate Runner Directories." When this button is clicked, for each file to be maintained, ModelMate checks the modification date of any existing copy of the file in each of the runner directories, and if the file in the master directory is newer or if the runner directory does not contain a file of that name, the file in the master directory is copied to the runner directory. If the checkbox labeled "Auto Populate Runner Directories" is checked, newer files are copied to the

runner directories as required whenever the user clicks the "Run UCODE" button on the left side of the main ModelMate window. Clicking "Create UCODE Input Files" in the main ModelMate window does not cause the runner directories to be populated with updated files. Click OK in the "UCODE Parallel-Processing Control" window to save data and close the window.

In some situations, a user may wish to use a subset of the runners defined in the "Parallel-Processing Runner Directories" window without deleting the information for the runner directories that are not to be used. For this situation, the control labeled "Number of Runners to Use in Next Parallel Run" in the "UCODE Parallel-Processing Control" window (fig. 17) can be used to set the number of runners to be used to any number up to the total number of defined runners. If the user chooses to set this control to a number less than the total number of defined runners, runners will be selected

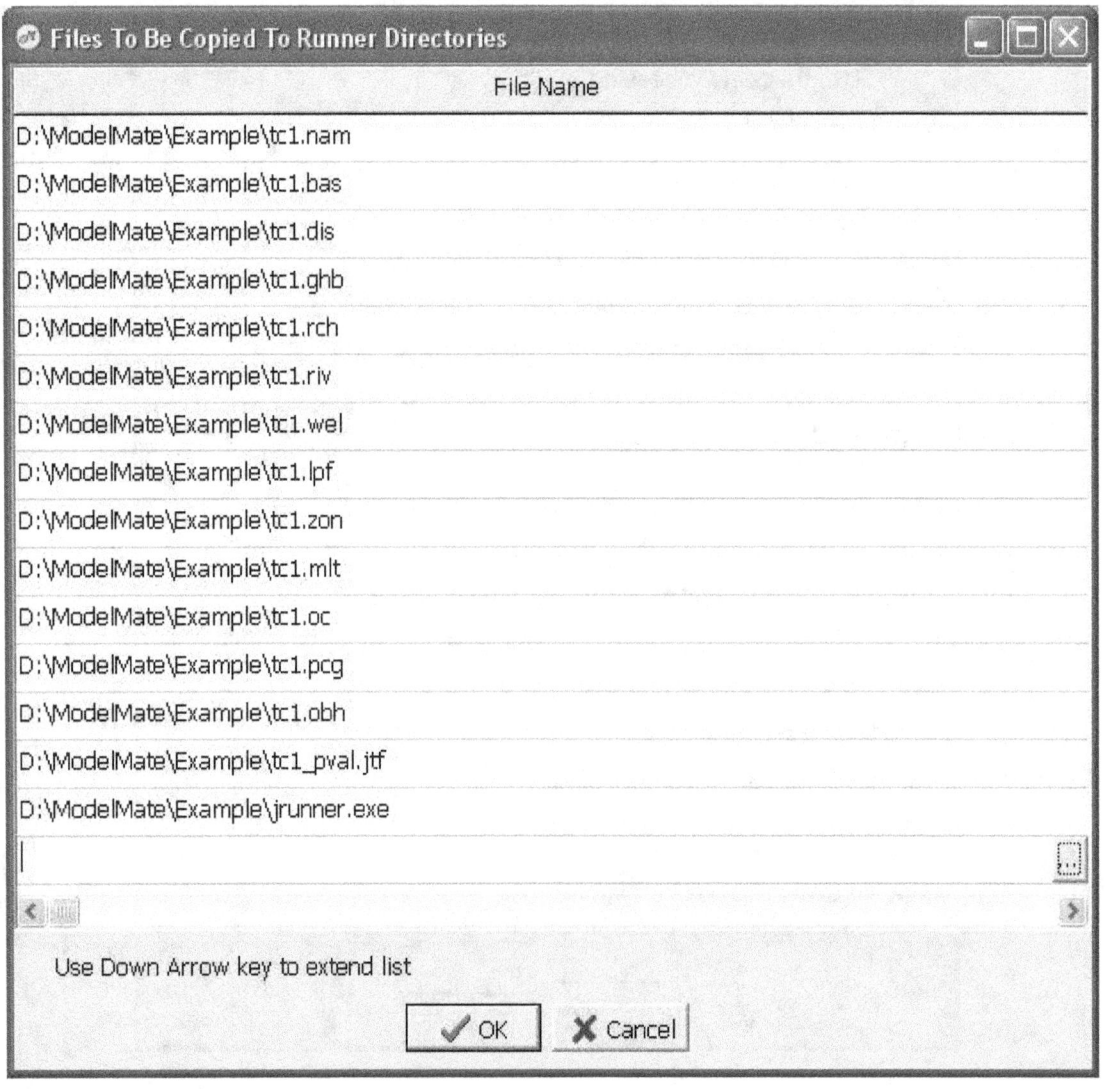

Figure 18.   Files To Be Copied To Runner Directories window.

based on their position in the "Parallel-Processing Runner Directories" window; one or more runners at the bottom of this list will not be used.

Starting and Stopping Runners

When UCODE is used for an analysis for which parallel processing is enabled, at least one runner needs to be running at the time UCODE is invoked; if no runner is running, UCODE will stop with an error message to this effect. If all runner directories are on remote computers, JRUNNER will need to be started manually in each runner directory. However, if at least one runner directory is on the computer running Model-Mate, the user may find it convenient to have ModelMate start the runner(s) on the local machine automatically just before UCODE is invoked. This way, UCODE can be started, and while the first model run is being made, the user can manually start JRUNNER in the runner directories on the remote computers. If the local computer has enough processors to host all desired runners, then ModelMate can be used to start all instances of JRUNNER. First, ensure that JRUNNER can be started in each of the runner directories. Then, if the "AutoStart Local Runners" checkbox on the "UCODE Parallel-Processing Control" window is checked, whenever parallel processing is to be enabled for a UCODE run in parameter-estimation mode or sensitivity-analysis mode, ModelMate will start JRUNNER in each of the local runner directories before invoking UCODE. ModelMate can distinguish runner directories on the local computer from runner directories on remote computers, so there is no need for additional input to define directories where JRUNNER should be started.

The checkbox labeled "AutoStop Runners" on the "UCODE Parallel-Processing Control" window is used to control the AutoStopRunners setting of the UCODE Parallel_Control input block. The effect of checking this checkbox is to cause all runners to stop at the end of a UCODE run; if the box is unchecked, runners reset themselves. If runners are reset, the user does not need to manually restart JRUNNER in the runner directories; however, the user would need to manually stop instances of JRUNNER if another UCODE execution is not to be made. If all runner directories are on the local computer, the most efficient use of processors is obtained by checking both the "AutoStart Local Runners" and "AutoStop Runners" checkboxes. When using runner directories on remote computers, personal experience can guide the use of these options. Instances of JRUNNER for the local runner directories also can be started from the ModelMate GUI if desired by selecting menu item **Parallel Processing|Start Local Runners**. This capability generally is not needed when ModelMate is used as a preprocessor for UCODE; however, advanced users of other model-analysis applications based on the JUPITER API may find this capability useful.

Postprocessing

UCODE is bundled with several postprocessor applications. These applications are summarized in table 4. RESIDUAL_ANALYSIS and RESIDUAL_ANALYSIS_ADV can be invoked from the **Postprocessing** menu (fig. 1). For instructions on running and interpreting results from these and the other four postprocessing applications, refer to Poeter and others (2005, p. 157–222) and Hill and Tiedeman (2007).

Table 4. Postprocessor applications included with UCODE.

[Postprocessor applications are documented in Poeter and others (2005, p. 157–222). Applications not supported by ModelMate can be executed outside of ModelMate as described by Poeter and others (2005)]

Application	Supported in initial version of ModelMate	Summary
RESIDUAL_ANALYSIS	Yes	Analysis of weighted residuals with respect to independence and normal distribution; leverage statistics.
RESIDUAL_ANALYSIS_ADV	Yes	Advanced statistical analysis of weighted residuals and preparation of data for graphical analyses.
LINEAR_UNCERTAINTY	No	Calculation of confidence and prediction intervals simulated with estimated parameter values and subject to the assumption of model linearity.
MODEL_LINEARITY	No	Quantification of model nonlinearity by modified Beale's measure
MODEL_LINEARITY_ADV	No	Calculation of total model nonlinearity, intrinsic model nonlinearity, and combined intrinsic model nonlinearity.
CORFAC_PLUS	No	Calculation of correction factors for confidence and prediction intervals to account for model nonlinearity.

Graphical Analysis

Many of the results generated by UCODE and its associated postprocessing applications are best interpreted by viewing them graphically (Hill and Tiedeman, 2007, p. 99–123). GW_Chart (Winston, 2000) is a graphing application specifically designed to work with files generated by UCODE and other applications to generate graphs that are useful in model analysis. After a successful UCODE run has been executed, ModelMate can be used to start GW_Chart by selecting menu item **Postprocessing|GW_Chart**. When invoked from ModelMate, GW_Chart always starts by producing a graph relating observation values to simulated values. Once GW_Chart has started, any of a number of graphs can be displayed by choosing an appropriate output file generated by UCODE or a related application. The extension of the file determines the type of graph that will be generated (table 5).

Table 5.   Graphs generated by GW_Chart using output generated by UCODE and related programs.

File name extension	Graph or analysis	Application used to generate file
_os	Observed compared to simulated values	UCODE
_ww	Weighted observed compared to weighted simulated values	UCODE
_ws	Weighted residuals compared to unweighted simulated values	UCODE
_nm	Normal probability graph of the weighted residuals	UCODE
_sc	Bar chart of composite scaled sensitivities	UCODE
_sd	Dimensionless scaled sensitivities for each observation by parameter	UCODE
_s1	One-percent scaled sensitivity for each observation by parameter	UCODE
_ss	Sum of squared, weighted residuals for each parameter-estimation iteration	UCODE
_pcc	Parameter correlation coefficient for parameter pairs	UCODE
_so	Bar chart of leverage for each observation and prior-information item	UCODE
_rd	Normal probability graph of random numbers	RESIDUAL_ANALYSIS
_rg	Normal probability graph of correlated random numbers	RESIDUAL_ANALYSIS
_rc	Bar chart of the Cook's D statistics for each observation	RESIDUAL_ANALYSIS
_rb	Bar chart of DFBeta statistics for each observation by parameter	RESIDUAL_ANALYSIS
_rdadv	Probability plot position compared to weighted residual	RESIDUAL_ANALYSIS_ADV

Advanced Topics

Use of UCODE Options Not Supported By ModelMate

ModelMate does not support all capabilities of UCODE. The following modes and other options of UCODE are not supported by ModelMate:

Modes not supported:

- Test Model Linearity

- Advanced Test Model Linearity

- Nonlinear Uncertainty

- Investigate Objective Function

Other options not supported:

- Correlation of observation errors (weight matrices)

- Derived observations

- Model-calculated derivatives of simulated values with respect to parameters

However, UCODE input generated by ModelMate can be edited with a text editor so that it will make a UCODE run that uses any of these options. First, use ModelMate to define settings and data for options that are supported. Then use the "Create UCODE Input Files" button (fig. 1) to create a UCODE main input file and batch file to start the UCODE run. Edit the UCODE main input file in a text editor and create additional input files as required, using instructions in Poeter and others (2005). Finally, use the batch file to start UCODE with the revised input.

Use of ModelMate with Applications Other Than UCODE

ModelMate can be useful when running applications other than UCODE that use the dispatcher/runner protocol of the JUPITER API to implement parallel processing. Any application that uses this protocol has a dispatcher program and a runner program, and it uses runner directories the same way as they are used in UCODE. ModelMate can be used to maintain files in runner directories that need to be identical to their counterparts in a master directory. To do this, start a new ModelMate project in a convenient location, define runner directories, and define a set of files in the master directory that need to be copied to the runner directories whenever the files are changed (see section titled "Maintenance of Runner Directories"). Save the ModelMate project. Whenever files in runner directories need to be updated to keep files identical, select menu item **Parallel Processing|Parallel Control** to open the "UCODE Parallel-Processing Control" window (fig. 17) and click the "Populate Runner Directories" button; this will copy files listed in the "Files to be Copied to Runner Directories" window (fig. 18) to the runner directories as needed.

ModelMate also can be useful in starting local runners for any application that uses the dispatcher/runner protocol of the JUPITER API. To do this, set up runner directories as described in the section titled "Maintenance of Runner Directories." Select one of the UCODE modes for which parallel processing is enabled (either sensitivity-analysis or parameter-estimation mode), and select **Parallel Processing|Start Local Runners** to start JRUNNER in all runner directories.

Use of ModelMate with Model Software Other Than MODFLOW-2005

In ModelMate, template and extraction-instruction files are generated automatically when parameters and observations are imported from MODFLOW-2005 input files. ModelMate does not provide a convenient way to generate template and instruction files for models based on other model software. If a model based on model software other than MODFLOW-2005 is to be analyzed, the user needs to generate template and instruction files in a text editor. Poeter and others (2005, p. 107–129) provide instructions for creating template and instruction files. When these files have been prepared, the user can use menu items **Model|Model Input and Template Files** and **Model|Model Output and Instruction Files** (appendix) to browse and select the various files as required. Then the user would need to enter names and data for parameters, observations, and(or) predictions interactively or by copying and pasting data from a spreadsheet program in the appropriate windows of the ModelMate GUI. In the "Model Settings" window (fig. 8) choose Generic as the model type; this setting disables controls related to MODFLOW-2005 in various windows.

Use of ModelMate with ModelMuse

ModelMuse (Winston, 2009) is a GUI designed to provide users with a graphical, interactive means for building a groundwater model. ModelMuse supports both MODFLOW-2005 and ModelMate. Users of ModelMuse can build a model based on MODFLOW-2005, export a ModelMate project file, and then use ModelMate for analysis and calibration of the model. If ModelMate is used to change parameter values by parameter estimation or manually, the new values can be imported back into ModelMuse for further simulations.

UCODE implements model analysis by manipulating parameter values, invoking the model, and extracting simulated values from model output. When building a model in ModelMuse that is to be analyzed by UCODE through the use of ModelMate, the user needs to define model input through the use of parameters. Many of the MODFLOW-2005 flow packages and stress packages support parameters, which can be defined in ModelMuse to control model input.

To maximize the ability of ModelMate to manipulate model input, define and use parameters in ModelMuse whenever possible and feasible according to the conceptual model. If specific model inputs are not expected to be involved in model analysis, the input can be defined by parameters but assigned as not adjustable in ModelMate. If the user later decides to include that model input in model analysis, it is much easier to include it if the model input is already defined by parameters. When ModelMuse exports MODFLOW-2005 input files for a model in which parameters are used, it writes a MODFLOW-2005 PVAL file and a template file to be used by UCODE to construct a PVAL file.

Model-simulated values that are to be used by ModelMate and UCODE need to be identified as either observations or predictions in ModelMuse. For a model that is to be used as the primary model in a ModelMate project, the user needs to define "Observation Type" as observations (ModelMuse menu item **Model|Observation Type**). For a model that is to be used as a predictive model in ModelMate, the user needs to define "Observation Type" as predictions in ModelMuse. In either case, after defining observations in ModelMuse, MODFLOW-2005 input files need to be written or rewritten by ModelMuse (generally by invoking a model run) because ModelMate will need to read the MODFLOW-2005 Observation Process input files in order to create instruction files.

When parameters and observations have been defined as desired in ModelMuse, a ModelMate project file can be exported from ModelMuse. First, provide a name for the ModelMate project file (ModelMuse menu item **Model|ModelMate Interface**). To export the ModelMate project file, use ModelMuse menu item **File|Export|Export or Update ModelMate File**. To verify that the export of the ModelMate file was successful, open the ModelMate file in ModelMate. The parameters and their values should be listed in the Parameters tab of the main ModelMate window. Data related to observations should be listed in the Observations window accessible from the Observations tab.

For convenience, ModelMate can be started from ModelMuse. To start ModelMate from ModelMuse, select ModelMuse menu item **File|Export|Export or Update ModelMate File**. A Save dialog box will appear in which the user selects the name of the ModelMate file. At the bottom of the Save dialog box is a checkbox labeled "Open with ModelMate". If this checkbox is checked, ModelMate will be started and will automatically open the ModelMate file that was just saved by ModelMuse.

ModelMate automatically creates instruction files for extracting simulated values from output files generated by the MODFLOW-2005 Observation Process only when observations are imported from MODFLOW-2005 input. When ModelMuse exports observations to a ModelMate file, these instruction files are not automatically created. However, instruction files readily can be created by ModelMate for observations defined in ModelMuse. To create instruction files, open the ModelMate project file in ModelMate and select ModelMate menu item **Model|Create Instruction Files**

For Observations Defined In ModelMuse. Doing so will create instruction files as required, and the model-output and instruction file names will be added to the ModelMate project. This step can be verified by selecting ModelMate menu item **Model|Model Output and Instruction Files**. If the model is a predictive model, instruction files for extracting model-simulated values as predictions can be generated similarly, by selecting ModelMate menu item **Model|Create Instruction Files For Predictions Defined in ModelMuse**.

Once the instruction file(s) have been created, any UCODE model analysis supported by ModelMate can be run. If a parameter-estimation run is made and if the optimum values are imported into ModelMate (see section titled "Running UCODE"), or if parameter values are changed manually, the user may want to transfer the new values to the ModelMuse project. To do so, save the ModelMate file and in ModelMuse select the **File|Import|ModelMate Values**. This option will allow the user to select a ModelMate project file—choose the file that was just saved. That file will be read, and data related to parameters and observations and selected other data will replace current values.

Renaming an Observation or Prediction

When MODFLOW-2005 input files are read to import information related to observations or predictions, instruction files are generated. These files contain observation and prediction names that correspond to the names listed in the "Observations" and "Predictions" windows of ModelMate. ModelMate is designed to allow the user to edit these names, but ModelMate does not have a way to modify the instruction files. The user is discouraged from editing observation and prediction names in ModelMate because doing so will cause UCODE to be unable to implement the application/model connection. However, instruction file(s) can be manually edited outside of ModelMate to use the revised observation or prediction names. If names must be changed, ensure that observation and prediction names in the instruction files are changed correspondingly.

Changing Directories For Process Models

In some situations it may be desirable to use ModelMate with alternative but similar models that use the same sets of parameters and observations. For example, a user may have two models that differ in spatial or time discretization but are otherwise similar. If the two models are stored in different directories, the user can switch models by using menu item **Model|Model Directories**. This will open a window titled "Model Directories," which can be used to change the directory used by ModelMate for the location of the main model input file. This directory is used when referring to various model-input and model-output files. The Model Directories window allows directories to be changed for the primary and the predictive model. Changing these directory names affects the directory that is referenced when:

- Building a UCODE input file;
- Searching for a MODFLOW name file;
- Building a PVAL template file;
- Displaying runner directories; and
- Displaying runner files.

Changing directory names does not copy any template, instruction, or other files.

Acknowledgments

Richard B. Winston of the USGS provided invaluable help related to design and programming of ModelMate. Eileen P. Poeter of the Colorado School of Mines and International Ground Water Modeling Center, and Mary C. Hill of the USGS provided numerous suggestions for improvements.

References Cited

Banta, E.R., Poeter, E.P., Doherty, J.E., and Hill, M.C., 2006, JUPITER—Joint Universal Parameter IdenTification and Evaluation of Reliability—An application programming interface (API) for model analysis: U.S. Geological Survey Techniques and Methods, bk. 6, chap. E1, 268 p., available at *http://pubs.er.usgs.gov/usgspubs/tm/tm6E1*.

Harbaugh, A.W., 2005, MODFLOW-2005, the U.S. Geological Survey modular ground-water model—the Ground-Water Flow Process: U.S. Geological Survey Techniques and Methods, bk. 6, chap. A16, variously paginated, available at *http://pubs.er.usgs.gov/usgspubs/tm/tm6A16*.

Harbaugh, A.W., 2010, A data-input program (MFI2005) for the U.S. Geological Survey modular groundwater model (MODFLOW-2005) and parameter estimation program (UCODE_2005): U.S. Geological Survey Open-File Report 2010-1057, 35 p., available at *http://water.usgs.gov/nrp/gwsoftware/mfi_2005/mfi_2005.html*.

Harbaugh, A.W., and Hill, M.C., 2009, OBS.pdf: File included in MODFLOW-2005 distribution, 32 p., available at *http://water.usgs.gov/nrp/gwsoftware/modflow2005/modflow2005.html*.

Hill, M.C., and Tiedeman, C.R., 2007, Effective groundwater model calibration, with analysis of data, sensitivities, predictions and uncertainty: New York, John Wiley & Sons, 464 p.

Parkhurst, D.L., and Appelo, C.A.J., 1999, User's guide to PHREEQC (version 2) a computer program for speciation, batch-reaction, one-dimensional transport, and inverse geochemical calculations: U.S. Geological Survey Water-Resources Investigations Report 99-4259, 312 p.

Poeter, E.P., Hill, M.C., Banta, E.R., Mehl, Steffen, and Christensen, Steen, 2005, UCODE_2005 and six other computer codes for universal sensitivity analysis, calibration, and uncertainty evaluation constructed using the JUPITER API: U.S. Geological Survey Techniques and Methods, bk. 6, chap. A11, revision of Feb. 10, 2008, 283 p., available at *http://pubs.er.usgs.gov/usgspubs/tm/tm6A11*.

Winston, R.B., 2000, Graphical user interface for MODFLOW, Version 4: U.S. Geological Survey Open-File Report 2000–315, 27 p., available at *http://water.usgs.gov/nrp/gwsoftware/mfgui4/modflow-gui.html*; the GW_Chart graphing program documented in this report is available at *http://water.usgs.gov/nrp/gwsoftware/GW_Chart/GW_Chart.html*.

Winston, R.B., 2009, ModelMuse—A graphical user interface for MODFLOW-2005 and PHAST: U.S. Geological Survey Techniques and Methods, 6–A29, 52 p., available at *http://water.usgs.gov/nrp/gwsoftware/ModelMuse/ModelMuse.html*.

Appendix.   Main Menu.

Menu [|Submenu]	Menu Item	Action
File	New Project	Clear current data and start a blank project. If current project contains unsaved changes, user is prompted to save the project.
	Open Project…	Open browse window to select and open an existing project file.
	Save Project	Save current project data to current file.
	Save Project As…	Open browse window to save current data in a new project file.
	Revert Project To Last Saved	Revert data to that stored in last saved project.
	Import	See menu item File|Import
	Exit ModelMate	Exit ModelMate; if current project contains unsaved changes, user is prompted to save the project.
File|Import	UCODE Main Input File As New Project…	Open browse window to select a UCODE input (*.in) file to be imported. If a file is selected, existing data are cleared and replaced by data from the imported file. If current project contains unsaved changes, user is prompted to save the project.
	Parameters And Observations From MODFLOW-2005…	Open browse window to select a MODFLOW-2005 name file. Parameters and observations read from files listed in the name file replace current parameters and observations.
	Parameters From MODFLOW-2005…	Open browse window to select a MODFLOW-2005 name file. Parameters read from files listed in the name file replace current parameters.
	Observations From MODFLOW-2005…	Open browse window to select a MODFLOW-2005 name file. Observations read from files listed in the name file replace current observations.
	MODFLOW2005 Observations As Predictions…	Open browse window to select a MODFLOW-2005 name file. Observations read from files listed in the name file replace current predictions.
	Optimized Parameters (_paopt file)…	Replace all parameter values with those listed in _paopt file generated by UCODE with base filename stored as output prefix in "UCODE Input and Output File Names" window.
Project	Project Name and Title…	Open "Project Name and Title" window.
	Program Locations…	Open "Program Locations" window.
UCODE	File Names…	Open "UCODE Input and Output File Names" window to provide filenames for UCODE input and output prefixes for UCODE output for primary and predictive process models.
	Settings…	Open "UCODE Settings" window to define model name, measurement units, UCODE functionality, and output options.
	Parameter-Estimation Settings…	Open "UCODE Parameter-Estimation Settings" window to define various controls related to parameter estimation.
	Derived Parameters…	Open "UCODE Derived Parameters" window to define parameters as functions of other parameters.
	Prior Information…	Open "Prior-Information Control" window, which allows access to "UCODE Linear Prior Information" window for defining prior information on parameters.

Appendix.   Main Menu.—Continued

Menu [ǀSubmenu]	Menu Item	Action
Model	Model Directories…	Open "Model Directories" window to define directories where input for process model(s) reside.
	Commands to Invoke Model…	Open "Model Commands" window to define commands used by UCODE for invoking a process model.
	Model Input and Template Files…	Open "Model Input and Template Files" window.
	Model Output and Instruction Files…	Open "Model Output and Instruction Files" window.
	Create Instruction Files For Observations Defined in ModelMuse	Create extraction-instruction file(s) to be used by UCODE to extract simulated equivalents to observations defined in ModelMuse.
	Predictive Model Input and Template Files…	Open "Predictive Model Input and Template Files" window.
	Predictive Model Output and Instruction Files…	Open "Predictive Model Output and Instruction Files" window.
	Create Instruction Files For Predictions Defined in ModelMuse	Create extraction-instruction file(s) to be used by UCODE to extract model-simulated values defined as predictions in ModelMuse.
	Model Settings…	Open "Model Settings" window.
View	UCODE Main Input File…	Open most recently generated UCODE input (.in) file with "ModelMate File Viewer."
	UCODE Main Output File…	Open the UCODE output (.#uout) file most recently generated by a UCODE run with "ModelMate File Viewer."
	Close All File View Windows	Close all windows that may be left open while the main ModelMate window is used. These include the "ModelMate File Viewer" windows.
	Refresh	Refresh the main ModelMate window.
Postprocessing	GW_Chart	Invoke GW_Chart (separate executable) with a UCODE data-exchange file of type dependent on selected mode.
	Residual_analysis	Invoke Residual_analysis (separate executable) with output most recently generated by UCODE.
	Residual_analysis_adv	Invoke Residual_analysis_adv (separate executable) with output most recently generated by UCODE.
Parallel Processing	Parallel Control…	Open "UCODE Parallel-Processing Control" window.
	Runner Directories…	Open "Parallel-Processing Runner Directories" window.
	Start Local Runners	Invoke jrunner.exe or runner.exe (separate executable) in one or more runner directories listed in "Parallel-Processing Runner Directories" window that reside on the local computer. Runner(s) will be started only if selected UCODE mode is "Sensitivity Analysis" or "Parameter Estimation." The number of runners started will be limited to the number shown in the "Number of Runners to Use in Next Parallel Run" box of the "UCODE Parallel-Processing Control" window.
Help	About ModelMate…	Open "About ModelMate" window.

www.ingramcontent.com/pod-product-compliance
Lightning Source LLC
Chambersburg PA
CBHW081410170526
45166CB00010B/3280